Sheep and Goat Practice 2

The *In Practice* Handbooks Series

Series Editors: Margaret Melling and Martin Alder

Current members of *In Practice* Editorial Board

I. D. Baker
A. L. Duncan
J. K. Dunn
H. A. O'Dair
T. J. Phillips
K. A. Urquhart (Chairman)

Titles in print:
Bovine Practice
Bovine Practice 2
Canine Practice
Canine Practice 2
Equine Practice 2
Equine Practice 3
Feline Practice
Feline Practice 2
Sheep and Goat Practice
Small Animal Practice

The *In Practice* Handbooks

Sheep and Goat Practice 2

Edited by M. Melling and M. Alder
Editors, *In Practice*

W. B. Saunders Company Ltd

LONDON   PHILADELPHIA   TORONTO   SYDNEY   TOKYO

W. B. Saunders   24–28 Oval Road
Company Ltd   London NW1 7DX

The Curtis Center
Independence Square West
Philadelphia, PA 19106-3399, USA

Harcourt Brace & Company
55 Horner Avenue
Toronto, Ontario M8Z 4X6, Canada

Harcourt Brace & Company, Australia
30–52 Smidmore Street
Marrickville
NSW 2204, Australia

Harcourt Brace & Company, Japan
Ichibancho Central Building
22-1 Ichibancho
Chiyoda-ku, Tokyo 102, Japan

A catalogue record for this book is available from the British Library

ISBN 0–7020–2330-2

Typeset by Photo·graphics, Honiton, Devon
Printed and bound in Hong Kong by Dah Hua Printing Press Co., Ltd

Contents

Contributors

T. Boundy, Tegfan, Montgomery, Powys SY15 6HW, UK

M.J. Clarkson, University of Liverpool, Department of Veterinary Clinical Science, Field Station, Leahurst, Neston, Merseyside L64 7TE, UK

D.D.S. Collie, c/o R. Caven, Carronvale, 38 Corstophine Bank Drive, Edinburgh EH12 8RN, UK

G.C. Coles, University of Bristol, Department of Veterinary Clinical Science, Langford House, Langford, Bristol BS18 7DU, UK

J.E. Cox, University of Liverpool, Department of Veterinary Clinical Science, Leahurst, Neston, South Wirral, Cheshire L64 7TE, UK

J.C. Hindson, Woodhouse Farm, Hatherleigh, Okehampton, Devon EX20 3LL, UK

B.D. Hosie, Scottish Agricultural College, Veterinary Investigation Centre, Bush Estate, Penicuik, Midlothian EH26 0QE, UK

S.S. Lloyd, University of Cambridge, Department of Clinical Veterinary Medicine, Madingley Road, Cambridge CB3 0ES, UK

D.J. Mellor, Professor and Head of Physiology and Anatomy, Massey University, Faculty of Veterinary Science, Private Bag 1/22, Palmerston North, New Zealand

V. Molony, Royal (Dick) School of Veterinary Studies, Department of Preclinical Veterinary Sciences, Summerhall, Edinburgh EH9 1QH, UK

P.R. Scott, Veterinary Field Station, Easter Bush, Roslin, Midlothian EH25 9RG, UK

M.A. Taylor, Parasitology Department, Central Veterinary Laboratory, New Haw, Addlestone KT15 3NB, UK

P. M. Taylor, University of Cambridge, Department of Clinical Veterinary Medicine, Madlingley Road, Cambridge CB3 0ES, UK

N.J. Watt, Royal (Dick) School of Veterinary Studies, Department of Veterinary Pathology, Summerhall, Edinburgh EH9 1QH, UK

A.C. Winter, Bryn Goleu Farm, Cornist Lane, Flint, Flintshire CH6 5RA, UK

G.N. Wood, The Old Manor Farmhouse, The High Street, Chippenham, Ely, Cambridge CB7 5PR, UK

Foreword

The importance of continuing professional development for the veterinarian has reinforced the value of the articles presented in *In Practice*. Originally published as a clinical supplement to *The Veterinary Record*, *In Practice* is now firmly established as a prime source of information for the experienced veterinary clinician and student. The articles, each specially commissioned from acknowledged authorities on their subject, are selected by an editorial board representing a broad spectrum of veterinary expertise with the aim of updating existing information or introducing new developments leading to changes in practice.

The convention of casting the articles in the form of 'opinionated reviews' with the emphasis, where appropriate, on differential diagnosis has proved extremely successful and continues to be a distinguishing feature of *In Practice*.

Republishing selected articles, each updated by the author, as *In Practice Handbooks* has proved very popular. For ease of reference, each handbook deals with a particular species or group of related animals. The present volume is one of the second series in what is likely to be a continuing set of *In Practice Handbook* titles.

CHAPTER 1

Routine Ram Examination

TERRY BOUNDY

INTRODUCTION

There is a good future for prime quality lamb, provided that the industry is able to offer the consumer an attractive and reasonably priced product throughout the year. This will only be achieved by an overall improvement in the quality of both the ewes and the rams in the national flock. The use of animals drawn from group breeding schemes and sire reference schemes with back fat scanning and early growth maturity records will no doubt go far in producing the required type of sires and dams in the not too distant future. Meanwhile, careful examination of the ram and the use of sires with good conformation and growth rate will improve the standard of lambs produced.

There is great variation in the fertility of rams. Unfortunately, there are no tests available that will accurately predict a ram's ability to get stock; the only real guarantee of a ram's breeding ability is the number and quality of the progeny produced under normal breeding conditions.

REGULAR FERTILITY EVALUATION

It is possible, however, to examine the different aspects of a ram's reproductive functions to estimate how closely it conforms to the generally accepted "normal" standards and to define any abnormality that may be present. In fact, one of the most useful services that a veterinary practice can offer a sheep owner is that of a regular fertility evaluation of the flock rams.

Unfortunately, the examination by a veterinarian of rams prior to service has had little support from the sheep industry. There is too much reliance on the use of a large number of rams to cover for those that may not be of a high standard. If, by chance, infertile or low fertility rams are dominant, they will always suppress the mating ability of a subordinate ram that could be of high fertility, with a consequent reduction in overall flock fertility. An annual examination prior to service should, therefore, be carried out by a veterinarian and the economic justification for such an examination explained to sheep owners.

A reproductive examination will not necessarily entail a collection of semen and its full examination. A careful physical examination will, in the majority of cases, be sufficient to indicate deficiencies and abnormalities that could cause reproductive failure (Figs 1.1 and 1.2). However, any ram suspected of infertility subsequent to a physical examination, e.g. orchitis, epididymitis, small soft testicles, spermatocele, etc., should have

Fig. 1.1 Suffolk terminal sire ram with excellent conformation.

Fig. 1.2 Kerry Hill ram with weak conformation of the hindlimbs, sickle hocks and long sloping pasterns.

a detailed examination carried out on its semen. A semen sample can be collected by means of an ejaculator, which is inserted into the rectum, or by persuading the ram to serve into an artificial vagina. The majority of rams do require some training if an artificial vagina is used.

The semen should be examined immediately under the microscope for density and motility. Stained specimens should be made and examined for live:dead ratio, abnormal spermatozoa and polymorphonuclear leucocytes. This examination will not determine degrees of fertility; it will simply show the status of the semen at the time of its collection which could be different to another examination a few days later. If groups of polymorphonuclear leucocytes are found in a semen smear a bacteriological and sensitivity test should be carried out on a sample of the semen.

PRE-PURCHASE AND PRE-SALE EXAMINATION

A pre-purchase, and for that matter, a pre-sale physical examination could help to avoid many of the after sale problems that appear to arise each year both from private purchases and from recognized ram sales – problems that so often end up as long drawn out expensive legal disputes.

Any discerning buyer of a valuable ram should always have a dependable examination of the animal carried out. Individual stud rams used for a single mating or for artificial insemination or embryo transfer should include a semen collection and evaluation. Such rams should also be tested for border disease and *Actinobacillus seminis*.

CERTIFICATION

All rams examined should be permanently identifiable. The owner should be supplied with the recognized certificate of veterinary examination of a ram intended for breeding. (Pads of these certificates can be obtained from TGS Subscriber Services, 6 Bourne Enterprise Centre, Wrotham Road, Borough Green, Kent TN15 8DG, telephone 01732 884023.)

PHYSICAL HEALTH AND CONDITION SCORE

The initial approach to any examination must include a full physical health check. Any physical or disease conditions which will prevent or reduce a ram's ability to serve normally will reduce the number of lambs it will sire. Any condition impairing olfactory, visual or auditory stimuli will act in a similar manner. The presence of internal parasites should not be overlooked. All newly purchased rams should be isolated and dosed forthwith with ivermectin to eliminate the possibility of anthelmintic resistant parasites spreading to a clean premises. Tick-borne fever must be considered as a possible cause of ram infertility.

It is important to assess the ability of the ram to remain reasonably sound during the whole of the breeding season under normal environmental conditions. A simple interdigital growth, for example, could grow rapidly in severe wet conditions and cripple a ram within a matter of days. It is important, therefore, to note any condition that should be corrected to enable the ram to perform to its maximum ability. It is equally important to recognize those conditions that could be transmitted to a ewe during the breeding season, such as orf and ulcerative balanoposthitis.

MANAGEMENTAL IMPLICATIONS

Condition score

Flock owners should aim to have their rams at the peak of reproductive efficiency at the time of joining the ewes. It is essential for rams to maintain condition at all times and not to become thin. Rams should be introduced at MLC condition score 3.5 to 4. It is important to stress to sheep owners that the two-month period prior to joining is critical because the development and maturation of sperm takes almost all of that time to complete.

Heat regulation

In the normal ram, the temperature of the scrotum is usually some 4–5°C lower than the rectal temperature when spermatogenic function is most efficient. Some breeds carry a heavy wool cover on the scrotum. If they are indoors and lying on a deep bed of straw for long periods sperm production can be reduced. It is better to advise the removal of all scrotal wool at shearing, not only to keep the testicles cool, but also to reduce the risk of *Dermatophilus congolensis* infection which is regularly encountered during routine ram examination causing scrotal dermatitis.

APPROACH TO THE EXAMINATION

A good knowledge of sheep breeds is, of course, essential, and a close study should be made of the National Sheep Association publication "British Sheep" before embarking on a physical examination.

Try initially to observe the rams as a group. An over the fence examination will generally indicate ailing rams.

Signs of good health are:

(1)   General alertness
(2)   Free movement
(3)   Freedom from lameness

(4) Close and uniform fleece
(5) Active feeding
(6) Rumination
(7) No visible wounds, abscesses or injuries.

Indicators of ill health include:

(1)   Listlessness
(2)   Abnormal posture and behaviour
(3)   Stiffness in movement
(4)   Persistent coughing or panting
(5)   Absence of cudding
(6)   Loss of condition
(7)   Persistent unthriftiness
(8)   Lameness
(9)   Diarrhoea
(10) Patchy loss of fleece
(11) Constant rubbing
(12) Separation from the group.

BODY EXAMINATION

Teeth

Longevity in sheep is closely related to the ability of their teeth to withstand wear and tear caused by grazing and feeding. Those sheep with incisor teeth closing firmly onto the upper dental pad are more likely to withstand the wear and tear of feeding on rough diets. A good bite, therefore, should always be considered superior to other forms of apposition (Fig. 1.3). Incisor teeth drift forward with age. Their premature loss can occur for many reasons and can even be associated with particular farms in an area.

Hard swellings of the lower jaw invading the incisor teeth, if not the result of an injury, could be caused by *Actinobacillus* species.

A jaw undershot to a moderate degree is a far better proposition than one in which all the incisor teeth rest on the front of the dental pad or completely in front of it. A jaw undershot by even 0.5 cm, say, would be unlikely to cause grazing or eating problems.

Fig. 1.3 (Left) A good bite. (Right) Incisor teeth showing no apposition with the dental pad. Brachygnathia to this extent is a serious conformational fault.

Teeth problems can often be the cause of marked unthrifti-ness in older rams. By running the fingers along the outside cutting edge of the upper molars, which can be felt through the cheek on either side, gross abnormal changes can be detected. Any swellings above the root area of the upper molar teeth or in the substance of the ramus of the lower jaw could indicate the site of a serious tooth problem.

Carefully examine the mouth using a gag designed for this purpose. At the same time, look for lesions of orf both inside the mouth and at the commissures of the lips.

Head and neck

There should be no persistent nasal discharge. Snoring rams must be suspected of pharyngeal or laryngeal injuries and should be rejected. Cases of laryngeal chondritis have been reported in Texel rams.

Pay particular attention to cracks at the base of horns and the presence of any infection. Head fly lesions can be severe and are a constant source of problems. Orf has been implicated with such head lesions.

The neck area is a common site for subcutaneous vaccine injections and some vaccines cause unsightly lumps no matter how carefully they are administered. Swellings in the submaxil-lary and parotid glands are referred to as "boils". They are not uncommon and persistently blow up, bursting and discharging greenish yellow pus. In this respect, true cases of caseous lymphadenitis must not be overlooked.

Head butting among fighting rams can be a cause of serious local damage and even death. Injuries involving the ears may produce acutely painful haematomas. Persistent ear irritation, with or without a discharge, should be investigated for the presence of psoroptic mange mites.

Painful sunburn or photosensitization can be common among light-skinned sheep.

Staphylococcal dermatitis (Fig. 1.4) involving the skin of the face and often extending to the eyelids is not uncommon among rams feeding from a confined trough space. Ringworm has also been diagnosed from facial dermatitis lesions.

Gid takes about 6 months to show CNS changes and is usually seen in lambs or yearlings.

All rams showing signs of "cud spilling" should be rejected.

Eyes

There should be no persistent ocular discharge. Examine the eyes for impaired vision. Blindness in one or both eyes could indicate a CNS condition. Cataracts have been diagnosed on a number of occasions. Keratoconjunctivitis, "pink eye", is often associated with groups of rams running together and can be a very painful condition. If neglected this condition can cause blindness.

Fig. 1.4 Staphylococcal dermatitis affecting the face and eyelids.

Entropion, a common hereditary fault in some breeds, is often treated surgically. Careful examination should be made for any residual signs of such surgery.

Fleece

The fleece should be carefully examined for signs of external parasites. *Dermatophilus congolensis*, causing a chronic and exudative infection with areas of crusting, can seriously affect dark-skinned breeds such as the Suffolk and the Clun Forest. The affected area always seems to be particularly prone to fly strike.

Patchy loss of wool with evidence of skin pruritis should be treated as a possible indicator of scrapie or sheep scab.

Brisket sores

Brisket sores can be a regular source of pain and infection during the summer, and affected rams will be down for most of the day to avoid fly irritation. These sores can be caused by ill-fitting sire harnesses but are usually precipitated by rough ground on an area where the skin is thin and unprotected by adequate wool cover.

Some vaccines are administered in the brisket area in order to avoid unsightly neck swellings and can cause painful swellings that interfere with general movement and with fitted sire harnesses.

Limbs and feet

The conformation of a ram's limbs is important; cow hocks, and straight hocks with long sloping pasterns are signs of weakness. The claws should be strong and well developed; severe turning in of the front or hind feet, splayed claws, shelly horn and corkscrew claws are all undesirable.

Interdigital growths are common to all breeds and have even been seen in yearling rams. They cause constant pain and lameness. They appear to develop as a soft fibrous growth from the

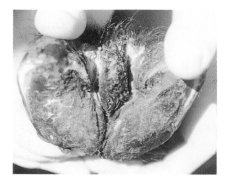

Fig. 1.5 Interdigital growth originating from the skin horn junction.

horn skin junction on the inside of the claw (Fig. 1.5); this may grow to grape-size and can give the appearance of a central growth. Once the external surface becomes eroded, infection occurs and in warm humid weather an interdigital growth may even become fly blown.

White line disease is probably the most commonly encountered cause of lameness (Fig. 1.6). The infection tracks up to the coronary band where it can invade neighbouring joints and tendon sheaths resulting in a severe purulent painful arthritis.

Footrot, with its characteristic manner of attacking the claw and its distinctive smell, is a condition that should not be overlooked. Strawberry footrot affects the lower limbs and can cause

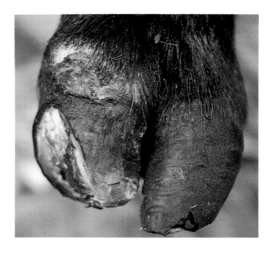

Fig. 1.6 White line disease.

slow healing, debilitating sores. The causal organism is *D. congolensis*, but it can be the result of a mixed infection with orf virus.

Scald appears as a mild inflammation of the skin between the claws with little or no extension to the horn of the claw.

Lameness is so common among sheep that it is only rarely that a flock is completely free of problems. Fortunately, most cases of lameness are of a simple nature and respond quickly to treatment if given early. Persistent and prolonged lameness in the 2-month period before service begins is one of the most common causes of reduced fertility. It can act indirectly by forcing the ram to spend long periods recumbent as well as directly by causing pain and suffering. Lameness reduces not only the ability of the ram to serve but also, if an infection is present, the active daily production of sperm. Post dipping lameness close to joining with ewes could have a disastrous effect on service results.

GENITAL EXAMINATION

To carry out an examination of the genitalia it is best to sit the ram on its tail in a well-lit area or place it into a "shepherd's chair" (see page 37). The testicles and scrotal area can be palpated in the standing position but it is not really possible to thoroughly examine the prepuce and penis in this position.

Prepuce and penis

Examine the prepuce closely looking for signs of injury. Ulcers at the junction of the skin and mucous membranes are common and can be associated with orf (Fig. 1.7). Note any abnormal discharge, particularly the presence of blood staining, that could indicate ulcerative balanoposthitis (Fig. 1.7). This condition can spread to ewes during service. It can lead to permanent scarring and adhesions in the ram, although pregnancy is rarely impaired. Homosexual practices are common among rams in a group and it is not uncommon to find the penis coated with faecal material.

In extending the penis (see Fig. 1.8), it is essential to ensure that it can pass through the preputial orifice and that there is

 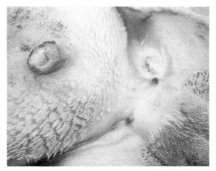

Fig. 1.7 (Left) Dermatitis and swelling of the prepuce, could be caused by orf. (Right) Gross swelling of the mucosa of the prepuce typical of ulcerative balanoposthitis.

Fig. 1.8 To extrude the penis grasp the sigmoid flexure firmly between finger and thumb, pushing upwards towards the prepuce and, at the same time, push the prepuce downward grasping it tightly. A further push from the sigmoid flexure area will completely extrude the penis.

no abnormal deviation in its onward direction. The urethral process should preferably be whole. Reduced fertility and even infertility may occur if it has been completely removed. Check the umbilical area for signs of herniation.

Scrotum and testicles

Start the examination of the scrotum (see Figs 1.9 and 1.10) and testicles high up against the body wall, lightly grasping the scrotum with both hands. Carefully examine the scrotum for the presence of boils, abscesses, shear cuts, dog bites, adhesions

Fig. 1.9 Palpation of the scrotum and contents assessing size, resilience and presence of abnormalities.

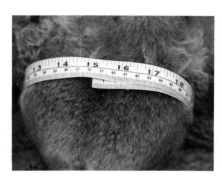

Fig. 1.10 Scrotal circumference correlates closely with testicular weight and, hence, sperm production. Ram testicular size varies with age, nutritional status, breed strain and time of year (it is low from December to July). Scrotal circumference of a group of 60 adult rams of various breeds, including Blue-faced Leicester, Texel, Charollais, Suffolk, Clun Forest, Kerry Hill and Bleu du Main, of a similar condition score and prior to joining with ewes, averaged 34–36 cm. A group of ram lambs of about 8 months old averaged 28–30 cm.

and dermatitis. A scrotal hernia will be indicated by a large soft mass between the testicles and the abdominal wall and there may also be some degree of hypoplasia on the affected side.

Squeeze each testicle gently and note the presence of any painful response indicating orchitis. The testicles should be more or less the same size and weight, heavy rather than light, and freely moveable within the scrotum. They should be plump and ovoid in shape and resilient; soft flabby testicles are undesirable. As the ram ages the testicles become soft and less resilient with the development of atrophy. A very hard firm testicle is pathological, e.g. orchitis.

In the general examination of groups of rams prior to joining with ewes, it can be assumed that testicles of even size, firm

elastic consistency and epididymal tails with a full resilient feel
are characteristic of the clinically normal ram and indicative of
normal testicular and epididymal function.

Abnormalities of testicular size and presentation

Hypoplasia

In assessing size of the testicles by palpation, one testicle may
appear to be smaller than the other. The condition of hypoplasia, sometimes associated with incomplete descent, is considered to be heritable (see Fig. 1.11). It is regularly seen in ram
lambs and has usually corrected itself by the time the lamb
becomes a yearling. Significant hypoplasia of both testicles
would be present when both testicles are about half the length
and half the diameter of those of other rams of a similar breed.
Rams with one testicle less than half the length and half the
diameter of the other should be condemned.

 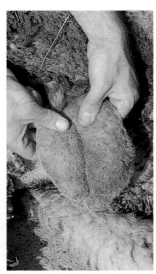

Fig. 1.11 (Left) Hernia
extending into the
scrotum with
hypoplasia of the
testicle of the
affected side.
(Right) Small poorly
shaped testicles with
a degree of
hypoplasia of one
testicle.

Fig. 1.12 Ram affected with acute painful orchitis standing alongside a ram with excellent conformation of the testicles.

Cryptorchidism and monorchidism

These conditions are both uncommon, but have been diagnosed (see Fig. 1.12).

Epididymitis

The epididymis should be large, rounded, resilient and free from any hard lumpy deformity. The significance of a hard enlarged tail is serious. Epididymitis is widespread throughout the world and mainly caused by *Brucella ovis*. This organism has not been found associated with rams in the UK; epididymitis has, however, been found in rams affected by *Histophilus ovis*, *Actinobacillus seminis* and *Escherichia coli*. Any abnormality of the epididymis including epididymal aplasia, should be carefully investigated (see Fig. 1.13).

Varicocele

On examining the spermatic cord at the head of the testicle, a firm lobulated mass is sometimes detected. It is probably a varicocele and could be heritable. A ram with this condition should not normally be used for service.

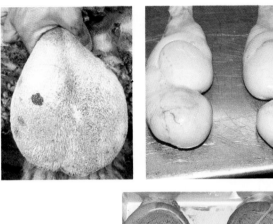

Fig. 1.13 (Upper left) Aplasia of the tail of the epididymis. (Upper right) Epididymitis and hypoplasia of both testicles. (Left) Epididymitis of the tail of the epididymis.

Spermatocele

Spontaneous and permanent hard swellings at the head of the testicle should be viewed with suspicion and rams so affected should not normally be used. In the acute stage these swellings can be painful and most certainly will reduce sperm production in the affected testicle. The heads of both the testicles pictured in Fig. 1.14 are affected with spermatoceles.

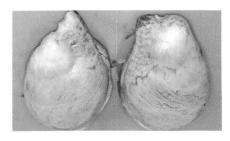

Fig. 1.14 Testicles affected with spermatoceles.

FACTORS AFFECTING SPERMATOGENESIS

A generally sound ram is one that exhibits no congenital, physical or genital abnormality which, in its progression, will cause a ram to be incapable of service. A fertile ram is one that is capable of service and is able to put ewes mated to him in lamb.

When joining with ewes all rams should be at a peak of reproductive efficiency. Ram lambs are capable of ejaculation as early as 18–20 weeks but great variation can occur in their ability to get ewes in lamb until they are 8 months of age, well grown and in good condition.

The penis–prepuce attachment is usually broken down by eight weeks of age and at around 20 weeks of age there is an alteration in the voice and the development of a very definite inguinal skin blush.

Breeding ability reaches its peak at between 2 and 5 years old but declines after 7–8 years. As rams age, a degree of atrophy of testicular tissue will occur. Both young and old rams, therefore, would be unpredictable choices for an intensive breeding programme.

Defects of an hereditary nature should be borne in mind and could include cryptorchidism, hypoplasia, aplasia, entropion, hernia and incisor teeth apposition.

LIBIDO

There is a great variation in the service activity of rams. Some, although appearing to be physically normal and capable of semen production, show little interest when introduced to a group of ewes in oestrus. Deficiency in libido is common among young rams but will usually rectify itself over a period of weeks.

Libido problems in older rams, and persistent problems in ram lambs, can be difficult areas on which to express an opinion.

FURTHER READING

British Sheep (1992) 8th edn. NSA, Malvern.

Boundy, T. (1985) *Care and Examination of Rams*, UVCE Tape Slide Programme. Royal Veterinary College, London.

Edgar, D. G. (1959) Examination of rams for fertility. *New Zealand Veterinary Journal* **7**, 61–63.

Edgar, D. G. (1963) The place of ram testing in the sheep industry. *New Zealand Veterinary Journal* **11**, 113–115.

Fraser, A. F. & Penman, J. N. (1971) A clinical study of ram infertilities in Scotland. *Veterinary Record* **89**, 154–158.

Galloway, D. B. (1966) Some aspects of reproductive wastage in rams. *Australian Veterinary Journal* **42**, 79–83.

Galloway, D. B. (1983) *Reproduction of the ram*. pp. 163–195. Post Graduate Committee in Veterinary Science Proceedings 67, Sydney, Australia.

Jensen, R., Swift, B. L. & Kimberlin, C. V. (1987) *Diseases of Sheep*. Lea & Febiger, Philadelphia.

Lindsay, D. R. & Pearce, D. T. (eds) (1984) Reproduction in sheep. Cambridge University Press, Cambridge.

Logue, D. & Greig, A. (1985) Infertility in the bull, ram and boar: 1. failure to mate. *In Practice* **7**, 185–191.

Logue, D. & Greig, A. (1986) Infertility in the bull, ram and boar: 2. infertility associated with normal service behaviour. *In Practice* **8**, 118–122.

Low, J. C. & Graham, M. M. (1985) *Histophilus ovis* epididymitis in a ram in the UK. *Veterinary Record* **117**, 64–65.

MacLaren, A. P. C. (1988) Ram fertility in south west Scotland. *British Veterinary Journal* **144**, 45–54.

Martin, W. B. & Aitken, I. D. (1991) *Diseases of Sheep*, Blackwell Scientific Publications, Oxford.

MLC Performance Recorded Rams (1985) MLC, Bletchley.

Waites, G. M. N. & Setchell, B. P. (1969) Physiology of the testis, epididymis and scrotum. In *Advances in Reproductive Physiology*, vol. 4, (ed., A. McLaren), pp. 1–63. Logos Press, London.

CHAPTER 2

Collection and Interpretation of Ram Semen Under General Practice Conditions

TERRY BOUNDY

INTRODUCTION

It is recognized among flock owners that the levels of fertility within a group of rams can vary significantly (Table 2.1). Unfortunately, there are no tests that will predict accurately a ram's ability to put ewes in lamb. The only guarantee of a ram's breeding ability is the number and the quality of the progeny produced under normal breeding conditions. It is possible, however, to examine the different aspects of a ram's reproductive function to estimate how closely it conforms to the generally accepted "normal" standard and to define any abnormality that may be present.

PHYSICAL EXAMINATION

A careful physical examination of the ram, preferably at least 2 months before joining with ewes, will generally be sufficient to indicate any deficiencies and abnormalities that could

Table 2.1 Factors interfering with sperm production.

Health
  Lameness
  Obesity
  Foot abscesses
  Infected interdigital growths
  Respiratory problems
  Headfly strike
  Scrotal dermatitis

Age and conformation
  Both young and old rams can be unreliable
  Heavy scrotal wool cover

Nutrition
  Long periods of over- or under-feeding
  Changes of feed

Management
  Changes of environment
  Confinement and overfeeding before sales
  Preparation for showing
  Prolonged periods of travelling and housing

potentially be a cause of reproductive failure; the finding of pathological conditions such as hypoplasia, orchitis, epididymitis or spermatocele, for instance, would indicate the need to examine the semen being produced by the ram. An examination at this time would enable advice to be given on such matters as condition score, feeding and simple treatable disease problems well in advance of tupping. If necessary, the owner would then have plenty of time to seek out replacements at late summer and autumn ram sales. Unfortunately, few owners request such an examination; practitioners, in the course of discussing flock health and management, would do well to encourage them to have their rams examined annually.

Some rams with apparently normal genitalia will show only a mild interest when joined with ewes in oestrus and this lack of libido may be permanent or temporary in nature. "Over the fence" examination of working rams should be stressed to flock owners, as a low libido ram working with a group of rams could easily slip through the net and remain an uneconomical passenger for a considerable period of time.

TESTICULAR WEIGHT AND SPERM PRODUCTION

Sperm production varies significantly throughout the year. At the height of season, the daily production is about 20 million spermatozoa per gram of testicular tissue. A ram, at this time, may ejaculate 15 or more times on each day spent with ewes in oestrus.

When blood testosterone levels fall, there is an accompanying and substantial reduction in the weight and size of the testicles (Table 2.2). Although sperm production does not entirely cease, it is considerably reduced. This factor must be borne in mind where an early lambing project is being considered; with artificial methods, ewes are easily brought into oestrus, but often little attention is directed towards the ram.

A rapid and substantial improvement in testicular size and sperm production can be achieved through the judicious use of rumen-undegradable protein in a ram's feed. Recently, the importance of zinc together with vitamin A has also been noted.

SEMEN COLLECTION

There are three methods by which semen can be collected from a ram. Irrespective of the method, rams should be taken away

Table 2.2 Variations in scrotal circumference among rams of a variety of breeds between March and June in one year.

Breed	Age (years)	Scrotal circumference (cm)	
		March	June
Suffolk	3	30	35
Suffolk	2	32	38
Suffolk	1	25	31
Bluefaced Leicester	1	31	36
Bluefaced Leicester	4	33	40
Kerry Hill	1	24	30
Texel	1	30	34
Jacob	2	32	35

from ewes preferably for 7 days and fed well before another collection is made.

Immediately subsequent to natural service, a collection can be made using a warm pipette inserted into the vagina of the ewe through a warm speculum. Because the normal volume of ejaculate that can be expected is only 1 to 2 ml, it is difficult to obtain a reasonable specimen by this method, and there is the added risk of contamination with female secretions and cells.

The best way of obtaining a truly representative semen sample is with the use of an artificial vagina. The method does have some problems in that only a few rams will mount and attempt to serve a ewe that is not in oestrus, and a period of training is required before a ram will ejaculate into an artificial vagina. This training period can take a number of days and depends a great deal on the patience and expertise of the operator. Occasionally, a ram will not accept this method of collection. Generally, however, it is possible to collect semen several times a day with an artificial vagina, and this method will form the basis of any artificial insemination programme.

The most practical method for the collection of semen for routine examination is by electrical stimulation from within the rectum with the use of an electroejaculator. This device consists of a probe which is inserted to its full length into the rectum, i.e., 12–15 cm in an adult ram. It is tilted down towards the pelvis and, in response to a small electrical charge from batteries within the rod, ejaculation is produced. Muscular contractions during manipulation will cause the ram some discomfort but there are no permanent ill effects. The use of the electroejaculator is confined to veterinarians and in each case it should be ensured that its use is really necessary – a physically normal ram, with no history or query of infertility, would not be a candidate for electroejaculation. Provided that some simple guidelines are followed (see below), the technique is not difficult to perform. The results, however, are entirely dependent on the patience, care and ability of the operator. There are some rams that will not produce a truly representative semen sample by this method, although with care and patience these are few in number. Ejaculation is an induced response, often without an erection, and the specimen will occasionally be contaminated with urine.

The infrequent and probable seasonal use of this equipment will necessitate removal and proper storage of batteries and a check for full power before use.

PROCEDURE FOR ELECTROEJACULATION

(1)   The area around the prepuce is shorn and the prepuce is wiped clean.

(2)   For a standing collection, the ram's head is held by an assistant. Further restraint is provided by the operator grasping the tail with one hand and standing with his or her leg in front of the ram's stifle area.

(3)   The probe is warmed and lubricated before being introduced into the rectum (if the rectum is packed with faeces, a soap and water enema is given prior to inserting the probe).

(4)   The ram is given a short while to acclimatize to the equipment (Fig. 2.1). During this period, the area around the brim of the pelvis is gently massaged with the probe. The operator maintains a firm grasp on the ram's tail throughout the procedure, to keep the probe (and him or herself) in position (Fig. 2.2 see also Fig. 2.3).

(5)   An electrical charge is applied for between 4 and 6 s and repeated after an interval of approximately 4 s. The majority of rams produce a specimen at this point (it may be necessary to wait a short while as the semen may not appear immediately after the charge ceases). If there is no response, repeat the process after a 10-min rest period. If semen is still not forthcoming,

Fig. 2.1 Preliminary conditioning of a ram to the presence of an electroejaculator.

Fig. 2.2 Operator and electroejaculator both positioned correctly for a standing collection.

Fig. 2.3 A semen collection may alternatively be obtained from a ram restrained on its side on a table or straw bales. Plenty of assistance is required as rams are very heavy. The advantage of this position is that it is easier to ensure that all the semen is collected in the container. It is also easier to obtain a sterile collection when the penis can be extruded, held with a short length of soft gauze, and the urethral process directed into a sterile collecting tube.

the procedure may either be attempted again later, or left for a week and repeated.

(6)   The ejaculate is collected by a second assistant into a warm container, e.g. a small plastic snappy bag, a graduated tube, a

plastic syringe container or even directly onto a warm microscope slide.

SEMEN EVALUATION

Spermatozoa are sensitive to cold shock, bright light, detergents, water, blood, disinfectants, metals, cigarette smoke and temperatures above 40°C. All equipment used for their examination should be kept at about 37°C. This can present difficulties in the field, although a baby's bottle heater or photographic hot plate will adequately overcome the problem.

Before making a semen collection, a microscope should be set up in a draught-free area. In order that the examination can be carried out as quickly as possible, the focus should be adjusted using × 100 magnification initially. A heated microscope stage and a water bath are useful, as is a phase contrast fitting, if possible.

SPERM CONCENTRATION

The concentration of spermatozoa is assessed on the visual appearance of the semen using the scoring system shown in Table 2.3.

The volume of semen can be measured using a graduated tube or pipette. The product of semen volume times sperm

Table 2.3 Visual assessment scale for sperm concentration.

Score	Approximate numbers of spermatozoa $\times 10^9$/ml
5 Thick creamy	4.5–6
4 Creamy	3.5–4.5
3 Thin creamy	2.5–3.5
2 Milky	1–2.5
1 Cloudy	0.3–1
0 Clear watery	Very few

concentration will give the total number of spermatozoa in each ejaculate. Good quality semen contains $3.5\text{--}6 \times 10^9$ spermatozoa/ml.

SPERM MOTILITY

Sperm motility is directly related to fertility and can be assessed by diluting the semen sample with warm 0.9% sodium chloride solution at 1:100, and observing individual spermatozoa under a cover glass at $\times$ 400 magnification. Over 50% of the spermatozoa should show forward progression. A green filter is a great help and, again, it is important that all equipment is kept at about 37°C.

WAVE MOTION

The wave motion within a semen sample is a function of the concentration of spermatozoa. Their activity can be scored as shown in Table 2.4.

The sample must be kept warm. Grades 4 and 5 only would be suitable for cervical artificial insemination. The test is, to a considerable degree, subjective.

Table 2.4 Spermatozoa activity scores.

Score	Description
5 Very good	Dense, rapidly moving waves. Individual sperm cannot be observed. 90% or more are active
4 Good	Vigorous moving waves, but not as rapid as score 5. 70–80% or more are active
3 Fair	Only small, slow moving waves. Individual sperm can be observed. 45–65% are active
2 Poor	No waves are formed. Some movement of sperm: 20–40% are active
1 Very poor	Very few spermatozoa show signs of life
0 Dead	All spermatozoa are motionless

SPERM MORPHOLOGY

The best way of studying the morphology of spermatozoa is to dilute a semen sample with formol saline and examine a thin wet film using phase contrast equipment. In practice, smears are usually prepared from a sample and stained and examined under an oil immersion objective. A number of staining methods are available, the most common being a mixture of 5% nigrosin and 1% eosin; 1% trypan blue is also recommended, particularly for those who are colour-blind to red. The stain should be prewarmed and the mixture (diluted 1:10) allowed to stand for a few minutes. The film must be made as thin as possible in order to differentiate single spermatozoa (Fig. 2.4); again, a green filter is considerable help. Two-hundred or more spermatozoa should be observed.

Abnormalities

All abnormalities should be noted and the percentage of abnormal sperm determined. All semen samples will contain some abnormal sperm – normal ram semen may contain up to 30% abnormal cells. A high percentage of primary or secondary abnormalities, and plasma containing a recognizable level of debris, could well influence a ram's fertility (Figs 2.5–2.7).

Primary abnormalities include detached heads, detached tails, abnormal head shapes, abaxial tails, double heads, double

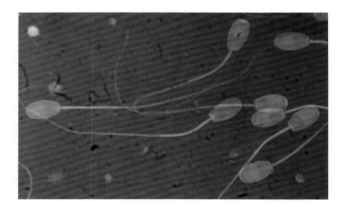

Fig. 2.4 Small group of physically normal spermatozoa. (× 1000).

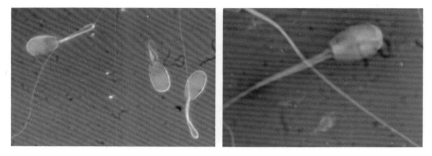

Fig. 2.5 (Left) Looped tails at about the point where the protoplasmic droplet is shed (× 1000). (Right) Double midpiece (× 1000).

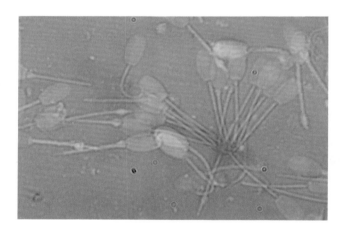

Fig. 2.6 Sperm agglutination with gross abnormalities; note the presence of many sperm carrying protoplasmic droplets on the midpiece and the level of debris in the plasma (× 1000).

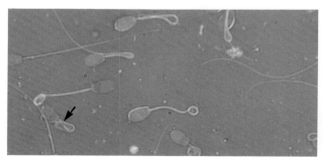

Fig. 2.7 Looped tails with protoplasmic droplets and dag defect tail (arrow) (× 600).

tails and dag tails (tails tightly coiled around the midpiece). These abnormalities develop in the course of spermatogenesis.

Secondary abnormalities are usually associated with the presence of a protoplasmic droplet remaining on the midpiece or in the loop of a coiled tail. These defects develop in the tail of the epididymis.

Tertiary abnormalities, such as reflection of the tail, occur after ejaculation and are probably caused by shock or injury in the preparation of the specimen.

Live:dead ratio

The ratio of live:dead spermatozoa should be at least 4:1. The two stains mentioned above have an affinity to select dead spermatozoa, and hence enable a fairly accurate estimate of the ratio to be obtained.

WHITE CELL INVOLVEMENT

Semen collected from any ram suspected of infertility should be examined for the presence of leucocytes, indicating inflammatory lesions of the genitalia and the accessory glands. A thin smear of semen is prepared and dried in air and stained with Leishman's stain (Fig. 2.8) or heat fixed and stained with

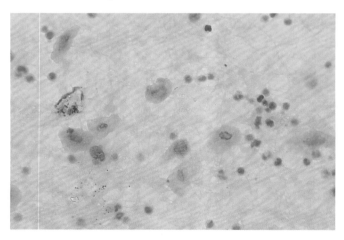

Fig. 2.8 Large numbers of polymorphonuclear leucocytes in a semen sample collected from a ram with epididymitis caused by *Histophilus ovis*. Leishman's stain (× 400).

methylene blue. Most of the cells seen will be preputial or urethral epithelium cells. The presence of large numbers of neutrophils suggests an inflammatory condition and this is likely to reduce fertility. Such samples should be submitted for bacteriological examination and sensitivity testing.

Testsimplets (Boehringer Mannheim, distributed by LabPak) are prestained slides that will stain spermatozoa for morphological interpretation and will also clearly show the presence of neutrophils. Spermatozoa stain densely, thus permitting easy recognition of the head, midpiece and tail. Neutrophils will stain with the cytoplasm containing colourless or reddishpurple fine granules similar to the standard Romanowsky technique. A rapid result can be obtained by taking a smear of semen with a swab, and rolling it into the prestained blue area. It is left for 2–3 min and then the background stain is washed off and the slide is allowed to dry. It is examined under a × 400 or oil immersion lens.

CONCLUSION

Semen examination will not determine degrees of fertility; it will simply assess the status of the semen at the time of collection. All stud rams for sale or export should, therefore, be examined by as many techniques as are practicable, to ensure at least a guarantee of apparent normality.

All ram testing should be carried out with great care and consideration for the welfare of the ram in question. It is of primary importance also that the ram is clearly identifiable. Too many rams are sold that are inadequate in respect of fertility and many of these problems could be resolved simply by improving the management and care prior to the sale. Young rams are particularly prone to fertility problems, but these generally sort themselves out as the rams settle into their new environment. Where a ram is the subject of an infertility examination subsequent to a sale, the auctioneer's rules should be carefully studied. The Suffolk Sheep Society, for example, will only accept an examination result for a semen sample that has been collected with the use of an artificial vagina.

The only certificate which should be provided after an examination is the official certificate of veterinary examination of a ram intended for breeding. (Pads of these certificates can be obtained from TGS Subscriber Services, 6 Bourne Enterprise Centre, Wrotham Road, Borough Green, Kent TN15 8DG, telephone 01732 884023.)

ACKNOWLEDGEMENTS

The author thanks Professor M. J. Clarkson and K. Smith for their helpful comments, and P. Jenkins for the preparation of the transparencies.

FURTHER READING

Barr, W. M. (1988) Assessment of ram fertility in hill flocks. *Proceedings of the Sheep Veterinary Society* **13**, 75–81.

Boundy, T. (1985) *Care and examination of rams*, UVCE Tape Slide Programme. Royal Veterinary College, London.

Cupps, P. T. (1991) *Reproduction in Domestic Animals*. 4th edn. Academic Press, London.

Evans, G. & Maxwell, W. M. S. (1987) *Salamons Artificial Insemination of Sheep and Goats*. Sydney University Press, Sydney.

Frazer, A. F. & Penman, J. N. (1971) A clinical study of ram infertility in Scotland. *Veterinary Record* **89**, 154–158.

Holt, N. (1987) Semen collection. pp. 25–28. Post Graduate Committee in Veterinary Science Proceedings 96, Sydney, Australia.

Hovell, G. J. R. (1988) Perspectives of ejaculation and semen quality in rams. *Proceedings of the Sheep Veterinary Society* **13**, 71–74.

Logue, D. & Greig, A. (1987) Infertility in the bull, ram and boar: collection and examination of semen. *In Practice* **9**, 167–170.

CHAPTER 3

Vasectomy in the Ram

TERRY BOUNDY AND JOHN COX

INTRODUCTION

Falling farm incomes, especially among hill and upland sheep farmers, together with increasing costs, have resulted in a reduced demand by stockmen for veterinary services. Much of the routine work traditionally carried out by veterinarians in sheep flocks has gradually been taken over by the shepherds; this has largely been made possible by the high standard of educational material now available through a variety of sources, and the sterling work carried out by veterinarians under the Agricultural Training Board. Increased emphasis on preventive medicine programmes and the introduction of proven management techniques would be of great benefit to both stock owners and veterinarians and would hopefully bring the two groups closer together. One very important aspect of management that is currently not being used to its full potential is the use of vasectomized rams to produce a compacted lambing period. In the UK, legislation states that vasectomy may only be carried out by a qualified veterinary surgeon.

SEXUAL BEHAVIOUR

Male sexual behaviour is the result of coordinated activity of the whole body and involves complex interactions between the environment, the sensory organs, central nervous system, endocrine system and reproductive organs. A sexually stimulating environment, under natural flock conditions, is the presence of a ewe in oestrus; ewes tend to seek out the ram and hence take some initiative in providing this environment.

The introduction of a ram into a flock will, within 30 min, increase the tonic luteinizing hormone (LH) pulses in the ewes, leading to a pre-ovulatory LH surge 27–36 h later. A "silent ovulation" (i.e., without behavioural oestrus) will occur within 3 or 4 days. In 40–60% of ewes, the corpus luteum formed after the silent ovulation is maintained for the normal duration, the first ovulation with overt oestrus occurring 18–20 days after the vasectomized ram is introduced. In the remaining ewes, the corpus luteum regresses by days 6–8, to be followed by a second ovulation when the corpus luteum is maintained for the normal duration to give a second peak of oestrus between days 22 and 24 (Fig. 3.1).

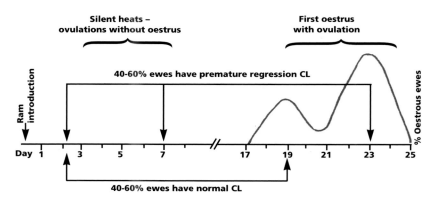

Fig. 3.1 Schematic diagram of the cyclical changes in ewes after the introduction of a ram early in the breeding season (K. C. Smith, personal communication (after Knight 1983)). CL, corpus luteum.

LIMITATIONS TO THE "RAM EFFECT"

There are, however, three limitations to this "ram effect":

(1) A proportion of the ewes in a flock may be cycling spontaneously and will not respond to the introduction of the ram.
(2) A proportion of those induced to ovulate may do so only once and will become anovular again, especially if the technique is used too far outside the breeding season.
(3) Not all anovular ewes will be induced to ovulate by the introduction of the ram, on account of natural biological variation.

PROCEDURE OF CHOICE

Vasectomy in the ram is a management technique used principally to produce a compacted lambing period. The ewe flock is isolated for at least 1 month from sight, smell and hearing of rams or wethers, after which a vasectomized ram is put with the flock for a period of 12 days. The ram is then removed and replaced by normal rams. Provided that the ewes were on the point of cycling, this results in a close compacted lambing, usually in two peak periods some 6–8 days apart.

Of the surgical techniques available for producing teaser rams, vasectomy is the procedure of choice, as it is in boars (Godke *et al.*, 1979) and stallions (Selway *et al.*, 1977). Other options have disadvantages associated with them, particularly as regards proven, long-term effectiveness. It is often stated that vasectomized animals lose their libido more quickly than normal entires. Management factors are likely to be important in determining how long a teaser animal will retain effectiveness and more work needs to be done in this area.

ALTERNATIVES TO VASECTOMY

Wethers treated with 150 mg of testosterone propionate three times at weekly intervals prior to joining will also act as rams and produce a "ram effect".

Two techniques that have been described as alternatives to vasectomy are surgical removal of the epididymal tail and blockage of the tube by injection of sclerosing agents. However, there remain doubts as to the long-term efficacy of these methods and they should therefore not be used.

SELECTION OF RAM

Because a vasectomized ram will be kept within a flock for a number of years, it is important that the right type of animal is chosen. It should be strong and healthy, with good incisor teeth occlusion, sound feet and no signs of any arthritic condition, or head or brisket sores. Young rams are preferable to older failed stud animals because they are relatively easy to operate on and there is less risk of them having previously acquired transmissible disease, such as border disease. The chosen ram should have had some sexual experience and well-developed genitalia. The penis and prepuce should be normal, with no signs of balanitis, and it should be ensured that the testicles and adnexa are free from any palpable pathological lesions such as epididymitis which might be caused by *Actinobacillus seminis* or *Histophilus ovis*.

If a ram is to be purchased for vasectomy, it should be chosen from a flock whose health status is known.

All vasectomized rams must be positively identified by tattoo, ear-tag or microchip (Fig. 3.2).

Fig. 3.2 Ram positively identified by a readily readable ear-tag.

SURGICAL TECHNIQUE

PREOPERATIVE MANAGEMENT

It is advisable that the ram is bedded on clean straw for at least 24 h and fasted for 12 h prior to surgery. Prolonged stalling should, however, be avoided, to reduce the risk of preoperative contamination. The scrotal wool and the abdominal wool around the neck of the scrotum should be shaved (not plucked) and the area should be washed and prepared routinely for surgery.

ANAESTHESIA AND RESTRAINT

General anaesthesia, although recommended, is not always practical; if used, the ram should be placed in dorsal recumbency (Fig 3.3). If sedation and local anaesthesia are to be employed for the procedure, the ram is best restrained in the sitting position – a shepherd's chair will facilitate surgery (Fig. 3.4). Good sedation is achieved with a 2% solution of xylazine, administered intramuscularly at a dose rate of 5 mg/100 kg bodyweight; this usually also gives excellent

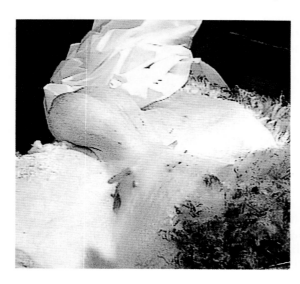

Fig. 3.3 Ram, in dorsal recumbency, prepared for surgery under general anaesthesia. Still taken from a video made by John Cox for the Veterinary Defence Society.

Fig. 3.4 Ram restrained in a shepherd's chair for surgery under sedation and local anaesthesia.

muscle relaxation. It should be noted, however, that this drug is not licensed in the UK for use in sheep. Local anaesthetic is carefully infiltrated just under the skin of the operation site on the cranial surface of the scrotum and into the fascia of the underlying tunic. There is a slight risk of injecting local anaesthetic into the pampiniform plexus, but this can be avoided by drawing back on the syringe prior to injection.

Excellent results have been reported by Sargison *et al.* (1993) using lumbosacral spinal analgesia and this new approach could become the method of choice. Their article should be consulted carefully before employing this procedure.

ANATOMY

Figures 3.5 and 3.6 show a cross-section through the neck of the scrotum and a caudal view of the scrotum. It should be noted that the vas lies medially within the scrotal neck with the cremaster muscle laterally and caudally.

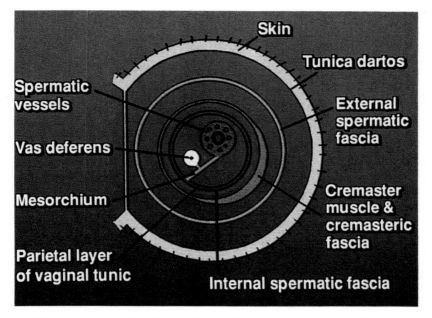

Fig. 3.5 Diagrammatic cross-section of the neck of (one half of) the ram's scrotum, showing the blood vessels and the various layers of tissue covering the vas deferens.

SURGERY

The vasectomy procedure is described below.

(1)   A vertical skin incision, about 4 cm long, is made on the cranial surface of the neck of the scrotum to the left of the mid-line over the neck of the left spermatic sac.

(2)   The left spermatic sac is freed by blunt dissection and exteriorized through the skin incision where it may be held in place with a pair of haemostats or tissue forceps.

(3)   The deferent duct lies medially within the spermatic sac so it is necessary to roll the sac outwards (Fig. 3.7). It is usually possible to identify the duct within the sac by its solid texture and by the presence of a small artery and vein which run close to it.

(4)   A nick is made in the vaginal tunic over the duct which then usually pops out; a spey hook helps to hold the duct out (Fig. 3.8). (The sac should not be incised until the duct is

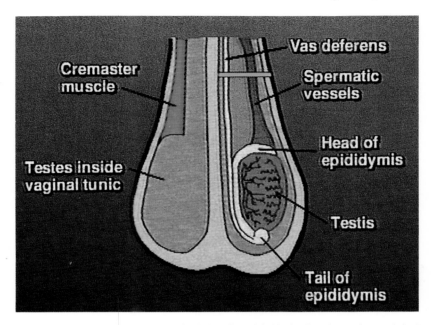

Fig. 3.6 Left and right testes of the ram as seen from behind. On the left side, the skin and external spermatic fascia have been removed to show the spermatic sac and cremaster muscle. On the right side, the inner layers of fascia and the parietal layer of the vaginal tunic have been removed to expose the vas deferens and blood vessels. The orange line indicates the level at which the cross-section (shown on page 39) was taken.

identified and held firmly below the point of incision. To incise the sac and then search for the duct is to invite a blood bath.) (5)   A portion of the duct is exteriorized and a length of not less than 3 cm (preferably more to reduce the risk of re-anastomosis) is resected (Fig. 3.9).

(6)   The ends of the vas are ligated and one end is anchored in the fatty tissue outside the vaginal tunic (Fig. 3.10).

(7)   The hole in the spermatic sac need not be sutured. Any significant dead space created by dissection of the sac is closed, but this is not usually necessary. The skin is sutured or stapled closed (Fig. 3.11).

(8)   The whole procedure, from skin incision to skin closure, is repeated to the right of the midline.

The pieces of tissue removed should be placed in fixative in separate pots (labelled left and right, as appropriate, and with the ram's identity tag) and sent for independent histological

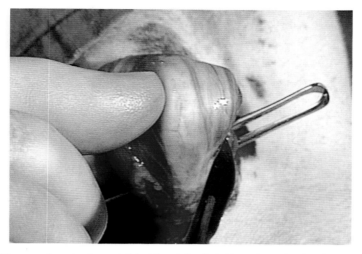

Fig. 3.7 The spermatic sac has been exteriorized through the skin incision and rotated to show the vas as a white structure within the sac.

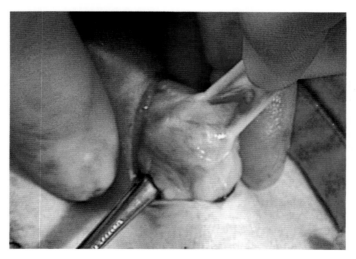

Fig. 3.8 A length of vas has been brought through a small nick in the fascia. The accompanying blood vessels can be clearly seen.

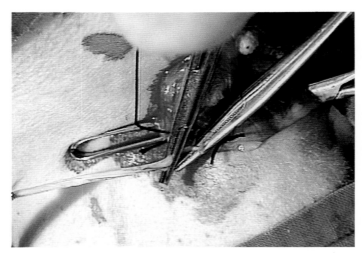

Fig. 3.9 A length of vas (not less than 3 cm) will be removed and the stumps ligated.

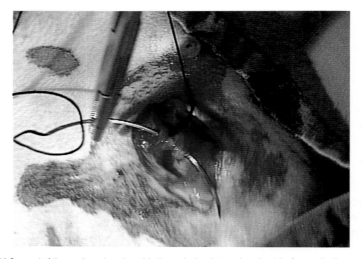

Fig. 3.10 One end of the vas is anchored outside the vaginal tunic to reduce the risk of recanalization.

analysis to confirm that they are, in fact, deferent duct. This will not only satisfy the surgeon that the correct tissue has been removed, but will also be useful evidence in the event of litigation in the future. As an immediate test, the contents of the excised duct may be squeezed on to a microscope slide and examined for the presence of sperm.

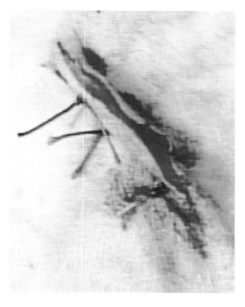

Fig. 3.11 The skin incision is sutured and the procedure repeated on the opposite side of the animal. Stills (Figs 3.7–3.11) taken from a video made by John Cox for the Veterinary Defence Society.

POSTOPERATIVE CARE

There should be no complications associated with the vasectomy procedure and, once the wounds have healed, the ram will be ready for work. Vasectomized rams may pass live sperm for the first few days. They may safely be introduced to a ewe flock after 1 week, although many operators prefer to leave them for up to 3 weeks and electroejaculate them prior to use. (Concern is often voiced regarding the welfare aspects of using an electroejaculator other than when absolutely necessary. It is worth bearing in mind that the ram does not ejaculate when the ejaculator is switched on, but in the interval between stimulation. Most rams will ejaculate after the second of two 4–6 s stimulations separated by a pause of approximately 4 s.)

Vasectomy should never be regarded as 100% reliable. It is important that the sheep owner is made aware of this and encouraged to contact the practice immediately should they have any doubts. It is certainly advisable that vasectomized rams are examined each year prior to the breeding season, ejaculated to check for the absence of sperm and the results recorded. Indeed, a proper and regular examination of all rams before they are put with the ewes can go a long way to saving both money and time – ram infertility is not uncommon and a cause of considerable loss to sheep owners each year.

OTHER USES FOR A VASECTOMIZED RAM

A vasectomized ram put with a group of ewes will indicate early signs of oestrus. The ewes may then be served in hand by selected rams. A vasectomized ram may also be used as a sweeper to indicate barren ewes at the end of a service period, although this method will not pick out all ewes which will not lamb because of losses occurring later in pregnancy.

ACKNOWLEDGEMENTS

The authors thank the Veterinary Defence Society for permission to use stills from a video film on vasectomy, made by John Cox on its behalf. The film is available on loan from the VDS at 4 Haig Court, Parkgate Industrial Estate, Knutsford, Cheshire WA16 8XZ.

REFERENCES AND FURTHER READING

Dunlop, A. A., Moule, G. R. & Southcott, W. M. (1963) Spermatozoa in the ejaculate of vasectomised rams. *Australian Veterinary Journal* **39**, 46–48.
Godke, R. A., Lambeth, V. A., Kreider, J. L. & Root, R. G. (1979) A simplified technique of vasectomy for heat-check boars. *Veterinary Medicine* **74**, 1027–1029.
Knight, T. W. (1983) Ram-induced stimulation of ovarian and oestrous activity in anoestrus ewes – a review. *Proceedings of the New Zealand Society of Animal Production* **43**, 7–11.
Pearce, D. T. & Oldham, C. M. (1984) The ram effect – its mechanism and application to the management of sheep. In *Reproduction in Sheep*. Cambridge University Press, Cambridge.
Sargison, N. D., Scott, P. R. & Woodman, M. P. (1983) The use of lumbo-sacral spinal analgesia in sheep. *Proceedings of the Sheep Veterinary Society* **17**, 177–179.

Selway, S. J., Kenney, R. M., Bergman, R. V., Greenhoff, G. R., Cooper, W. L. & Ganjam, V. K. (1977) Field technique for vasectomy. *Proceedings of the Annual Convention of the American Association of Equine Practitioners* **23**, 355–361.

Weaver, A. D. (1967) Vasectomy in the ram. *Veterinarian* **4**, 155–159.

CHAPTER 4

Meeting Colostrum Needs of Newborn Lambs

DAVID MELLOR

INTRODUCTION

It is well known that early intakes of colostrum by lambs during the first day after birth improve survival rates and it is obvious that measures which cater for orphans and lambs from ewes with insufficient colostrum would be beneficial. Nevertheless, problems arise because farmers often do not know how much colostrum a lamb needs nor the quantities of colostrum which are available for fostered lambs or which are obtainable by milking; nor are they aware that large volumes of colostrum can be obtained from ewes quickly and effectively by hand milking. Practical advice in these problem areas is now available.

Generally, the quantities of colostrum lambs require and ewes produce, and the ease of acquiring it by hand milking, have all been greatly underestimated.

WHAT IS COLOSTRUM?

Colostrum usually accumulates in the udder during the final few days of pregnancy. It is also produced during the first 12–

24 h after birth, but is diluted progressively as milk production increases. The transition from colostrum to milk is gradual and is accompanied by marked decreases in the concentrations of antibodies and sodium, and by increases in potassium and lactose. The usual definition of colostrum emphasizes a high immunoglobulin content, but in this chapter the word colostrum will be used simply to mean 'early milk', i.e. all the exocrine secretions produced by the udder during the first 24 h after birth.

The functions of colostrum are:

(1)   It contains nutrients which fuel heat production and thereby helps to prevent hypothermia.
(2)   It contains substances, including growth factors, which promote gut growth and differentiation especially during the first 24–48 h after birth, thereby helping to establish the enteral route as the lamb's sole source of nutrients.
(3)   It contains immunoglobulins, some of which line the gut wall while the rest are absorbed into the bloodstream – these help to prevent infections.

Unitl recently, we had placed such emphasis on the high immunoglobulin content of colostrum that we had inadvertently underestimated the importance of colostrum in providing metabolic fuels for heat production.

HOW MUCH DOES A LAMB NEED?

It is important to distinguish between how much colostrum a lamb can drink, how much it does drink and how much it needs to drink.

Lambs fed to appetite by bottle every 2 h can drink about 270 ml colostrum/kg bodyweight (equivalent to 1080 ml for a 4 kg lamb) during the first 18 h after birth. This represents the maximum intake.

The amount of colostrum a lamb drinks depends on the quantities available and on the success of sucking. Colostrum availability is affected by breed, ewe nutrition during late pregnancy and the number of lambs born, whereas sucking success depends on the early establishment of a good ewe–lamb bond and on competition between litter mates. Obviously, colostrum intake can vary in different lambs between zero and the maximum. The amounts of colostrum a lamb needs depend mainly on how much fuel it requires for heat production. Therefore, any factor which increases heat production increases the colostrum requirement. During bad weather (cold, wind and, or, rain) the lamb must produce more heat to avoid hypothermia, so its requirement for colostrum increases.

In order to avoid hypothermia, lambs born in average field conditions (0–10°C, wind, rain) need about 210 ml colostrum/kg bodyweight during the first 18 h, while intakes of about 180 ml/kg would be adequate in housed animals (still, dry air at 2–10°C). These figures must be multiplied by the lamb's weight (kg) in order to estimate the total volume of colostrum required during the first 18 h after birth (Table 4.1).

These quantities will normally also be sufficient to protect lambs against gut infections which cause watery mouth or diarrhoea because 200 ml of colostrum usually contain enough immunoglobulins for this purpose. Colostral immunoglobulins have specificities against particular pathogens and if that specificity is absent, protection will not be afforded. When lambs are to be reared on commercially produced milk substitutes which contain no immunoglobulins, it is essential to provide them with at least 200 ml colostrum during the first day to reduce the risk of gut and other infections during early life.

These colostrum requirements represent the total intakes during the first 18 h after birth. When feeding orphan or sick lambs by stomach tube it is important to avoid excessive stomach distension, so it is not wise to give more than 50 ml/kg on each

Table 4.1 Colostrum requirements during the first 18 h after birth.

Lamb weight (kg)	Colostrum requirements (ml)	
	Indoors[*] (2–10°C, no wind, dry)	Outdoors[+] (0–10°C, wind, rain)
2.5	450	525
4.0	720	840
5.5	990	1155

[*] 180 ml/kg
[+] 210 ml/kg.

occasion. It is convenient to use 50-ml syringes to deliver the colostrum as this saves times and permits the simple instruction that at each feed a lamb should receive one syringe-full for each kg of its weight. It is necessary to feed each lamb four or five times during the first 18 h after birth.

HOW MUCH IS AVAILABLE?

When first confronted with the notion that lambs require 180–210 ml colostrum/kg during the first 18 h after birth, some farmers are very sceptical about whether ewes can produce such large volumes of colostrum. The reasons given include the following: the udder of sheep is small (compared with those of goats and dairy cows); such farmers can only extract quite small volumes by hand milking (usually without using a teat lubricant and with the ewe upended between their legs); the volumes of colostrum thought to be necessary to prevent infections are much smaller. However, there is now no doubt that ewes which are healthy and have been fed adequately during late pregnancy can produce the volumes of colostrum indicated above.

The amounts of colostrum available depend mainly on ewe nutrition during late pregnancy and the number of lambs carried. There are also breed differences.

In well-fed ewes (body condition scores of 3 to 4), large volumes of colostrum accumulate in the udder just before birth and large volumes continue to be produced during the first 18 h after birth. Conversely, underfed ewes (condition scores 1.5–2) have very little colostrum in their udders at birth and it takes several hours for colostrum production to increase. The overall colostrum production by underfed ewes is usually about half that of well-fed ewes during the first 18 h, but in some underfed ewes no colostrum is produced.

When ewes are well fed (see Table 4.2), more colostrum is produced as the number of lambs carried increases in order to provide enough for the extra lamb(s). In underfed ewes, however, the presence of twins or triplets is often associated with the production of less colostrum, because a heavy lamb burden usually demands more nutrients than the ewe can supply when feed quality and/or availability are low. In such cases the pres-

Table 4.2 Average values for the total colostrum requirements of single and twin lambs and the total yields of colostrum during the first 18 h after birth.

Breed	Nutrition	Number of lambs	Colostrum requirement (ml)		Colostrum yield (ml)
			Indoors	Outdoors	
Scottish blackface	Well fed	1	860	1000	1805
		2	1420	1660	2080
	Underfed	2	1160	1350	990
Suffolk	Well fed	1	930	1090	2340
		2	1510	1780	2835

ence of two or more lambs can cause undernutrition in ewes which would otherwise be eating sufficient, or it can transform what would be "moderate" undernutrition into "severe" undernutrition.

Colostrum supply is therefore more likely to be inadequate when the ewe is in poor body condition at lambing, and particularly when she has two or more lambs. These principles apply for all breeds.

SPARE COLOSTRUM AND FOSTERING

Well-fed ewes (condition score 3–4) with single or twin lambs usually produce more colostrum than their lambs need; those with single lambs produce enough for a second lamb, particularly if the colostrum requirement is reduced by fostering indoors. Conversely, underfed ewes with twins usually have less colostrum than their lambs require and it would be necessary to foster at least one lamb in each case. Alternatively, lambs may be reared by hand or, in some cases, kept with their ewes and supplemented by stomach tube until copious milk secretion is established. In both situations it is advantageous to provide them with colostrum during the first day after birth. Accordingly, the routine collection and freezing of surplus colostrum from suitable donor ewes is recommended, beginning at the start of each lambing period.

Donor ewes generally fall into one of two categories: healthy ewes which have lost their lambs and which are not used for fostering or well-fed ewes with one or two lambs – in the latter care must be taken to ensure that the lambs at foot receive enough colostrum. Ewes which have aborted as a result of infection or for unknown reasons should not be used.

MILKING DONOR EWES

(1)   Keep the ewe standing normally – do *not* upend her.
(2)   If she has lambs keep them where she can see and lick them.
(3)   Immediately before milking on each occasion, inject into a jugular vein 5 i.u. oxytocin (e.g. Oxytocin-5, Intervet). Alternatively, inject 10–15 i.u. oxytocin intramuscularly. The intravenous route is preferable in skilled hands because let-down then takes 30 s as opposed to 2–3 min and about 20% more colostrum can be obtained.
(4)   Lubricate the teats with jelly (e.g. K-Y Lubricating Jelly, Johnson & Johnson), liquid paraffin, Vaseline (Johnson & Johnson) or saliva. Milk the ewe gently.
(5)   Start milking the ewe, using a large mug or a plastic jug as a container. Use a second container, placed safely out of kicking distance, to hold the colostrum as it is obtained. Within 30 s or 3 min of correctly performed injections the colostrum will flow freely. If it does not flow there is none there.
(6)   Milk ewes without lambs up to three times during the first 18–24 h after birth. Although extra colostrum can be obtained by milking the ewes more often, the additional volume is not worth the effort involved. Convenient milking times are within the first 2 h and at 10–14 h and 18–24 h after birth. Remove residual lubricant from the teats after milking.
(7)   When milking well-fed ewes with lambs at foot make sure the lambs have enough colostrum by: milking only one side of the udder, or not emptying the udder at each milking time, or milking out both sides once only, but not before 6 h after birth and only after checking that the lamb(s) have sucked adequately.
(8)   The adequacy of colostrum intakes in lambs at foot may be tested by offering them colostrum or milk from a bottle. Pro-

vided that they are normothermic and otherwise healthy, refusal to drink indicates that their intakes are sufficient. It is important to remember, however, that the sucking drive disappears when a lamb's rectal temperature decreases below about 37°C, so hypothermia should be considered when thin lambs refuse to drink.

VOLUMES OBTAINED BY HAND MILKING

Table 4.3 shows the volumes of colostrum which can be obtained from Scottish blackface and Suffolk ewes, with no lambs at foot, which are milked straight after an intravenous oxytocin injection at 1 h, 10 h and 18 h after birth. The 1-h yield mainly represents the colostrum which accumulates before birth, and the 10-h and 18-h yields represent the colostrum which is produced after birth. Clearly, oxytocin injections allow the collection of large volumes of colostrum from a small number of ewes. Similar volumes are obtainable from other British breeds and from the Australian merino.

Milking ewes in this way will not have harmful effects on subsequent milk yields; if anything it will increase them because it ensures that the udder is emptied completely several times during the first day after birth, and this promotes milk production.

STORAGE AND USE

It is convenient to keep the colostrum in 0.2 and 1.0 l containers. These can be clean yoghurt pots, soft-drink bottles or milk bottles, but whatever container is used it must be sealed before storage. The colostrum may be used during the same or a subsequent lambing period because when it is deep frozen (−20°C) ewe colostrum does not deteriorate for at least 1 year. Before feeding, the colostrum should be heated slowly, in warm *not* hot water, to about 37°C. It is not known whether thawing colostrum in a microwave oven damages the immunoglobulins, so it may be wise to avoid the practice. Unused, thawed colostrum

Table 4.3 Average volumes of colostrum obtained by hand milking.

Breed	Nutrition	Number of lambs	Total lamb weight (kg)	Colostrum production (ml)			
				1 h[*]	10 h	18 h	Total
Scottish blackface	Well fed	1	4.78	600[+]	575	630	1805
		2	7.90	715	675	690	2080
	Underfed	2	6.43	160	375	455	990
Suffolk	Well fed	1	5.17	650	860	830	2340
		2	8.37	865	1120	850	2835

[*]The 1-h yield represents mainly the prenatal accumulation and the subsequent yields the postnatal production of colostrum.
[+]Yields can show marked differences between animals even when they are on the same plane of nutrition, so these volumes should be taken as a guide only.

should not be refrozen, but can be kept refrigerated (4°C) for up to 48 h, after which it should be discarded.

SUBSTITUTES FOR EWE COLOSTRUM

Spare ewe colostrum may not be available when needed, but both goat and cow colostrum can be frozen and used as substitutes when necessary.

Goat colostrum has a similar composition to ewe colostrum and may contain the bonus of clostridial antibodies if the goats have been vaccinated. However, the feeding of lambs with colostrum from goats infected with caprine arthritis-encephalitis (CAE) which is similar to maedi-visna disease in sheep, may pose a threat to the lambs. If possible, therefore, goats used for feeding lambs should be tested for CAE and used only if the test is negative.

Cow colostrum may be used if ewe colostrum is not available, but with two provisos. First, the volumes required should be increased by about 20–40% because cow colostrum contains less nutrients than ewe colostrum. It is important to note any signs of hunger in the lambs and adjust their colostrum intakes accordingly. Second, cow colostrum should only be fed on the first one or two days after birth, because prolonged feeding of cow colostrum to lambs can cause haemolytic disease in some circumstances. Cow colostrum contains antibodies to the many bacteria and viruses to which the cow has been exposed and these apparently help prevent infections in lambs fed cow colostrum. In addition, treating cows with clostridial vaccines will allow their colostrum to provide lambs with appropriate protective antibodies.

CONCLUSIONS

(1)   Colostrum provides fuel for heat production and immunoglobulins for protection against infections.
(2)   In order to meet its fuel needs, the newborn lamb requires large quantities of colostrum during the first 18 h after birth.

(3)   The quantities needed during the first 18 h vary according to the environmental conditions and the birth weight of the lamb.

Indoors (2–10°C, no wind, dry): 180 ml/kg bodyweight;

Outdoors (0–10°C, wind, rain): 210 ml/kg bodyweight.

The total amount required by each lamb during the first 18 h after birth varies between 450 ml and 1200 ml.

(4)   Well-fed ewes have plenty of colostrum for their lambs. Underfed ewes will not have enough colostrum, particularly if they have two or more lambs.

(5)   Well-fed ewes with one lamb have enough colostrum for a second lamb.

(6)   Well-fed ewes that have lost their lamb(s) are suitable either for fostering or hand milking, provided that the ewes have not aborted.

(7)   Giving oxytocin by injection, either intravenously or intra-muscularly, just before milking the ewe greatly improves the yield of colostrum obtained. Let-down is faster and more complete with the intravenous route.

(8)   Using this method, milking ewes three times during the first day after birth will yield between 850 ml and 2400 ml of colostrum.

(9)   Banking colostrum obtained from suitable donor ewes early in the lambing season is recommended. When deep frozen it will keep without deteriorating for at least 1 year.

FURTHER READING

Barlow, R. M., Gardiner, A. C., Angus, K. W., Gilmour, J. S., Mellor, D. J., Cutherbertson, J. C., Newlands, G. & Thompson, R. (1987) A clinical, bio-chemical and pathological study of perinatal lambs in a commercial flock. *Veterinary Record* **120**, 357–362.

Eales, F. A. (1982) Detection and treatment of hypothermia in newborn lambs. *In Practice* **4**, 266–269.

Eales, F. A., Gilmour, J. S., Barlow, R. M. & Small, J. (1982) The causes of hypothermia in 89 lambs. *Veterinary Record* **110**, 118–120.

Eales, F. A., Small, J. & Gilmour, J. S. (1982) The resuscitation of hypothermic newborn lambs. *Veterinary Record* **110**, 121–123.

Mellor, D. J. (1983) Nutritional and placental determinants of fetal growth rate in sheep and consequences for the newborn lamb. *British Veterinary Journal* **139**, 307–324.

Mellor, D. J. (1988) Integration of perinatal events, pathophysiological changes and consequences for the newborn lamb. *British Veterinary Journal* **144**, 552–569.

Mellor, D. J. & Cockburn, F. (1986) A comparison of energy metabolism in the newborn infant, piglet and lamb. *Quarterly Journal of Experimental Physiology* **71**, 361–379.

CHAPTER 5

Anaemia in Lambs and Kids Caused by Feeding Cow Colostrum

AGNES WINTER AND MICHAEL CLARKSON

INTRODUCTION

Cow colostrum can be fed to lambs and goat kids as a source of both immunoglobulins and energy where a shortage of homologous colostrum exists or when attempting to control the spread of colostrum-transmitted diseases such as maedi and caprine arthritis-encephalitis. The rationale for using bovine colostrum is that immunoglobulins and other beneficial substances in colostrum are not species specific and are absorbed into the circulation during the period of intestinal permeability. Although the protection afforded is unlikely to be as good as that gained from colostrum from the homologous species, it is much better than providing no colostrum at all. Thus, in most cases, there are only beneficial effects to be gained from this practice.

However, anaemia in young lambs (Fig. 5.1) resulting from the feeding of cow colostrum was reported in the Netherlands in 1982 (Franken *et al.*) and in the UK in 1983 (Stubbings *et al.*); a similar condition has been reported in kids in France (Perrin and Polack, 1988). Various agents of the disease in both lambs and kids have been investigated by Winter (1989, 1990).

Outbreaks are linked to specific cows, so that on many farms where cow colostrum is used the problem is never experienced.

Fig. 5.1 Normal (right) and anaemic (left) lambs. Note the striking contrast in the colour of the conjunctivae.

The proportion of cows whose colostrum has this effect is not known, but is likely to be fairly low otherwise more incidents would have been reported. However, on an individual farm where colostrum from one of these "toxic" cows is used, a significant number of lambs or kids can be affected and some are likely to die, resulting in a major economic loss. It is important, therefore, to be able to diagnose the condition and to be able to offer advice on treatment and prevention.

AETIOLOGY

The anaemia is a result of absorbed bovine IgG becoming attached to the surface of the circulating red blood cells and their precursors in the bone marrow. These cells are then recognized as "foreign" by the reticuloendothelial system and are destroyed at an increased rate. Because the red cell precursors are destroyed as well as circulating red cells, the affected animal may be unable to manufacture sufficient replacement cells at what is already a time of great demand upon the bone marrow, as cells containing fetal haemoglobin become replaced by cells containing adult haemoglobin.

The reason why this phenomenon occurs in only some lambs and kids fed with the colostrum of only a small proportion of cows has not yet been established, although a blood group incompatibility is one possibility. Factors such as amount fed, time after birth of feeding and whether a previous feed of homologous colostrum has been given will all influence the amount of bovine IgG taken into the circulation of any individual animal and are likely to account for variations in the severity of clinical signs seen in an outbreak. In the case of a particularly toxic cow, one feed of as little as 50 ml may be sufficient to induce anaemia in some of the animals to which it is fed, although a greater quantity will generally be fed.

CLINICAL SIGNS

The destruction of red cells does not usually occur as quickly as in cases of neonatal isoerythrolysis in other species, and affected lambs and kids may appear normal to the casual observer until the packed cell volume (PCV) has fallen to about 0.10 l/l (normal range: 0.27–0.35 l/l). Clinical illness is usually evident at about 5–12 days old. Affected animals show the following clinical signs:

(1) Cease to suck milk voluntarily
(2) Obvious signs of weakness and lethargy
(3) White mucous membranes
(4) Markedly increased heart and respiratory rates
(5) May be observed to drink water if accessible
(6) May show pica
(7) Jaundice (occasionally)
(8) (Haematuria is not normally present).

The PCV usually continues to fall as low as 0.03–0.04 l/l and the lambs or kids usually die within a few days unless (and sometimes even if) treated. Animals may sometimes simply be found dead, particularly if they have not been observed sufficiently carefully in the first week or two after birth. The provision of milk with the aid of a stomach tube may prolong the animal's life, although any procedure which involves handling a young animal in such a weak state may precipitate death. Once the condition has been recognized within a flock, it is often possible to pick out less severely affected animals by examination of the conjunctivae. If necessary, a blood sample

Fig. 5.2 Post mortem appearance of an anaemic lamb. Note the extreme pallor of the whole carcase.

can be taken for measurement of PCV. Any animal with a PCV of less than 0.20 l/l should be viewed with suspicion and a decision made as to whether treatment is warranted. Some mildly affected animals will recover without treatment, but this should be given if obvious clinical signs of anaemia are present, or if the affected animal is particularly valuable.

POST MORTEM APPEARANCE

The carcase appears very pale (Fig. 5.2) and only a small amount of watery blood is present after removal of the major organs (Fig. 5.3). Slight jaundice may occasionally be present.

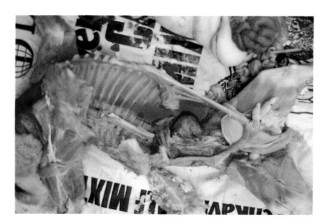

Fig. 5.3 Note the small amount of watery blood remaining when the internal organs have been removed.

The digestive tract will usually be empty of milk, unless a stomach tube has been used for feeding before death. Strands of hay and straw may be present in the stomach.

The liver and kidneys are usually of normal, although pale, appearance. Bone marrow is creamy white or grey, in contrast to the bright red coloration of a normal animal (Fig. 5.4).

DIAGNOSIS

Diagnosis should present little problem, since there are few other conditions which produce a similar clinical picture. It is based on:

(1)   A history of having been fed cow colostrum during the first day of life
(2)   The clinical signs of gross anaemia
(3)   Characteristic post mortem appearance of the bone marrow.

Other conditions which could be considered in the differential diagnosis of anaemia in very young lambs are:

(1)   Internal haemorrhage resulting from trauma at or following birth

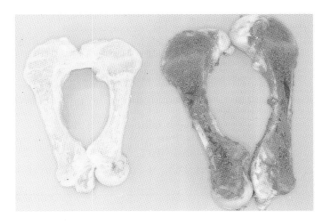

Fig. 5.4 Contrasting appearance of bone marrow of normal (red) and anaemic (pale) lambs. This is the most striking post mortem feature in these cases.

(2) Occasional outbreaks of heavy *Taenia hydatigena* infection causing liver haemorrhage (although this would usually affect slightly older lambs)
(3) Iron deficiency anaemia, which has recently been reported in housed flocks.

If further confirmation is required in the live animal, a direct antiglobulin (Coombs') test using a commercial preparation of antibovine IgG may be carried out on a sample of washed red cells. Titres of 1:640–1:2560 may be found in severe cases. This is also a useful test in determining the likelihood of recovery of an individual affected animal, or in deciding whether further treatment is required, as a marked fall in titre generally precedes recovery. However, this test is unlikely to be available in most practice laboratories, nor will it be an economic proposition in most cases.

TREATMENT

There are several practical problems in considering treatment once a diagnosis has been made. These are:

(1) The numbers of animals likely to be involved

Table 5.1 Treatment checklist.

A Severely anaemic (PCV < 0.10), not feeding	Blood i.v., i.p., or combination Corticosteroid for 4 days, check PCV, continue if not rising Iron dextran Antibiotic cover
B Moderately anaemic (PCV 0.10 to 0.15), feeding	Corticosteroid for 4 days then check PCV Continue if falling, stop if rising Iron–dextran Antibiotic cover
C Known to have had affected colostrum	Check PCV, if falls below 0.15 treat as (B)

i.v., intravenously; i.p., intraperitoneally.

(2) The relatively small economic value of each (although lambs or kids with a potentially high individual value may be found in some pedigree flocks)
(3) The presence within the flock of other lambs (kids) which may have been fed the same colostrum but have not yet developed clinical disease. Unless there is a good recording system within the flock, these may not be readily identifiable.

The methods of treatment that are available are intravenous blood transfusion, intraperitoneal administration of blood, corticosteroid administration and iron supplementation (see Table 5.1). When corticosteroids are given, antibiotic cover should also be provided. Anabolic steroids and various vitamin and mineral preparations have also been used, but there is no evidence that they are beneficial.

INTRAVENOUS BLOOD TRANSFUSION

Blood is collected, taking sterile precautions, into a commercial bloodpack. Crossmatching of blood is not a practical possibility; therefore, it is taken from any animal, although it is probably sensible to collect from the dam, if available. Blood can then be given to the recipient via a cannula with a 21 G butterfly needle placed in the cephalic vein. This vein is usually as easy to locate as in dogs and cats, except that it tends to run rather more diagonally across the front of the forelimb. The volume which can be given by this method is greater than by the other methods described, and 120 ml given over a period of 30 min has been suggested (Wain and Redpath, 1985).

If a commercial bloodpack is not available, or its use is not an economic proposition, blood may be collected, with sterile precautions, into a large (50–60 ml) syringe containing an anti-coagulant (50 i.u. heparin or 1 ml of 3.8% sodium citrate solution per 10 ml blood). The blood can then be injected slowly, i.e. over 10–15 min, via a cannula or needle, preferably into the cephalic vein. With patience, 10–15 ml per kg bodyweight may be successfully administered by this method, but it is very easy to inject too quickly and cause death as a result of overloading the circulatory system in animals in such a fragile state. Administration of blood can be carried out only once with each animal, because repeat transfusions carry a risk of a transfusion reaction occurring. It is important, therefore, to give the maximum amount which can be safely administered by the chosen route.

INTRAPERITONEAL ADMINISTRATION OF BLOOD

In outbreaks in which a number of animals of low value are affected, or where time to perform intravenous administration is limited, it is possible to give blood via the intraperitoneal route with some hope of success because the cells are taken up into the lymphatic system and are returned to the circulation within 24–48 h.

Blood is collected into a large syringe containing anticoagulant as above. The site for injection into the abdomen is as described by Eales *et al.* (1982) for intraperitoneal injection of glucose to hypothermic lambs, i.e., 1.5 cm lateral and 2.5 cm caudal to the umbilicus, with the lamb or kid held vertically by the front legs (Fig. 5.5). The site should be clipped and sterilized before injection, for which a 18 G 2.5 cm needle is suitable. The volume which can be given by this method is about 15 ml/kg bodyweight, and the procedure is quick and is well tolerated even by kids.

A further possibility is to give a small amount (5 ml/kg) of blood by slow intravenous injection and a further 15 ml/kg by intraperitoneal injection.

CORTICOSTEROID THERAPY

Corticosteroid therapy is an important component of therapy, both for severely anaemic animals in conjunction with administration of blood, and for less severely affected animals which do not yet require blood therapy. Corticosteroids probably slow down both the destruction of the IgG tagged cells and of transfused cells. A course of four or more injections at daily intervals, beginning with 1–2 mg/kg and reducing progressively, is suggested.

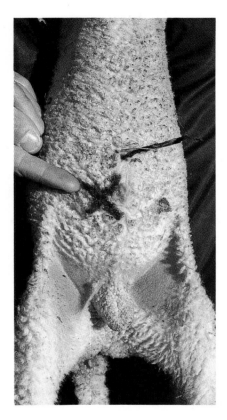

Fig. 5.5 The site for intraperitoneal injection should be clipped and sterilized. The injection is made with the needle angled backwards at 45° to the abdominal wall.

ADMINISTRATION OF IRON

Although young ruminants have generally not been thought to require supplementation with iron, there is evidence that housed animals may suffer from iron deficiency anaemia. Therefore, administration of iron–dextran is likely to be beneficial. A single dose of 200–300 mg is suggested.

PREVENTION

The ideal preventive measure would be a "cow-side" test to determine whether the colostrum is safe to feed to lambs or kids. However, no such test currently exists.

In the Netherlands, the technique of double immunodiffusion in agarose gel (whey extracted by ultracentrifugation is tested against a panel of sheep sera) has been employed with apparent success to screen samples of colostrum, reducing the incidence of anaemia outbreaks markedly, though not entirely eliminating them (Bernadina and Franken, 1985). This test is outside the scope of the practice laboratory and is not offered by commercial laboratories in the UK.

A method which assesses the effect of colostral whey extracted with rennet, or serum taken from potential colostrum donor cows, on a panel of sheep red cells in the presence of complement has also been tested (Winter, 1990), with some success, but again this test is not generally available.

Until a quick, convenient and affordable test is available, the best advice to farmers wishing to feed cow colostrum is to save it from several cows (preferably at least four) and to pool before feeding, so that any potential toxic factor is diluted to a degree that is unlikely to cause problems. If there are practical difficulties of insufficient cows calving at the same time, individual collections can be frozen until sufficient are stored, then thawed, mixed and refrozen in small batches. This extra thawing and refreezing is unlikely to seriously damage the IgG content if performed carefully and without overheating during thawing.

If colostrum from single cows is to be fed, the identity of the donor cow should be recorded and lambs to which it has been fed should be individually identified, for example, by colour spray marking. Failure to record creates difficulties in identifying animals which may need treatment before it is too late, and has also hindered research in this field because the identity of the "toxic" cow is not known.

ACKNOWLEDGEMENTS

The authors thank the Whitley Animal Protection Trust which generously funded the work on which this chapter is based and Dr F. A. Eales for supplying the intraperitoneal injection site figure.

REFERENCES

Bernadina, W. E. & Franken, P. (1985) A single method for the demonstration of factors in bovine colostrum capable of causing anaemia in lambs reared free from maedi on bovine colostrum. *Veterinary Immunology and Immunopathology* **10**, 297–303.

Eales, F. A., Small, J. & Gilmour, J. S. (1982) Resuscitation of hypothermic lambs. *Veterinary Record* **10**, 121–123.

Franken, P., Bernadina, W. E., Konig, C. D. W., Elving, L., van den Ingh, T. S. G. A. M. & van Dijk, S. (1982) Anemie bij zwoegervrij opgefokte schapelammeren (Anaemia in maedi-free artificially reared lambs). *Tijdschrift voor Diergeneeskunde* **107**, 583–585.

Perrin, G. & Polack, B. (1988) Bovine colostrum warning (letter). *Veterinary Record* **122**, 240.

Stubbings, D. P., Gibbons, D. F. & Tindall, J. R. (1983) Feeding cows' colostrum to newborn lambs (letter). *Veterinary Record* **112**, 88–89.

Wain, E. B. & Redpath, J. A. (1985) Blood transfusion as a treatment of anaemia in lambs (letter). *Veterinary Record* **116**, 527.

Winter, A. C. (1989) Anaemia in lambs caused by feeding cow colostrum. *Proceedings of the Sheep Veterinary Society* **14**, 49–53.

Winter, A. C. (1990) The feeding of cow colostrum to neonatal lambs and kids. PhD Thesis, University of Liverpool.

CHAPTER 6

Welfare Aspects of Castration and Tail Docking of Lambs

GRAHAM WOOD AND VINCENT MOLONY

INTRODUCTION

The reasons for castration and tail docking have been regularly debated; debate continues in the light of new knowledge, changes in husbandry and as attitudes change.

The main reason lambs are castrated is to prevent indiscriminate breeding and thus to maintain genetic control of breeding stock. Castration also helps to prevent mating of young females not in an adequate physical state to undergo pregnancy or parturition, and can reduce transmission of venereal diseases. Sexually based behaviour is eliminated and this reduces the risk of injuries both to other sheep and to the animal itself. It is widely believed that the conformation and quality of the carcase produced is improved by castration because there is an increase in fat deposition and therefore less collagen in the muscle and any possibility of taint is avoided. Downgrading of carcases of uncastrated lambs occurs under some circumstances and producers are unwilling to risk such losses.

The main arguments against castration are that lambs can reach an acceptable carcase size of better quality before they reach puberty and become sexually active. Uncastrated lambs generally grow faster, more efficiently and produce leaner carcases. In addition, there are no checks to growth resulting from acute and chronic pain, stress or infection and there are none of the accompanying welfare problems. Many types of husbandry do not, however, permit rapid growth and castration may be the only way to avoid the problems outlined above.

Tail docking is carried out mainly to prevent disease, the most serious problem being the accumulation of faeces and urine on the wool of the tail and hindquarters which can lead to bad hygiene or, at worst, fly strike and death. The welfare costs of these conditions could be great but no quantitative studies are available to confirm such speculation. The simplest and most effective preventive treatment is considered to be removal of part of the tail but more extensive treatment is required for some breeds in some countries where the operation of mulesing is carried out. (Mulesing is the removal of strips of skin from the perineal area of lambs so as to increase the area of woolless skin and confer a lower susceptibility to fly strike.)

The main arguments against tail docking are that problems of soiling and fly strike do not occur in all flocks, that the act of tail docking can be inhumane and that it can lead to further welfare problems such as chronic pain and sepsis. In addition, the lamb is deprived of the normal use of its tail for removal of flies and for protection and insulation of its hindquarters.

METHODS OF CASTRATION AND TAIL DOCKING

Despite various studies comparing the three methods commonly used, none has been adopted as being better than the others in all circumstances. These methods are:

(1) Surgical or "open"
(2) Ischaemic, by application of tight rubber rings
(3) Ischaemic, by application of a Burdizzo clamp.

Chemical and immunological methods have been investigated experimentally but are not used in the UK for lambs.

SURGICAL CASTRATION

The testes, epididymis and the spermatic cord to above the pampiniform plexus are normally removed through one or two incisions which expose the testis on each side. Haemorrhage from the spermatic and epididymal arteries is reduced or eliminated by using emasculators, cautery, ligatures or most commonly in lambs by twisting and tearing of the spermatic cords. Bleeding from the scrotal skin is usually limited in extent. Accumulation of blood is reduced if the incision(s) reaches the most dependent parts of the scrotum and if the dartos muscle contracts vigorously after castration.

This method is unequivocal, the testes are removed and can provide neither a source of pain nor a focus for infection, although the incisions in the scrotum can.

SURGICAL TAIL DOCKING

Part of the tail is removed by cutting or twisting using a knife or shears. Haemorrhage is limited by the tearing action, by using thermal cautery which may be incorporated in the shears, or a Burdizzo clamp may be applied before cutting off the tail. Without such attention to haemostasis the wound is less aesthetically acceptable, although there may be no significant physiopathological effects from the haemorrhage.

ISCHAEMIC CASTRATION WITH TIGHT RUBBER RINGS

Ischaemic castration with rubber rings relies on occlusion of the arterial supply and venous drainage of the testes, which run in the spermatic cord, by applying pressure around the neck of the scrotum (see Fig. 6.1). The rubber rings commonly used are approximately 15 mm outside diameter and 5 mm inside diameter in their unstretched state. They are stretched over the scrotum with a special instrument (elastrator) before being released with both testes trapped in the distal scrotum. The method is effective, quick and easy to carry out but, without anaesthesia, its use is restricted to the first week of life (Fig. 6.2).

G.N. Wood and V. Molony

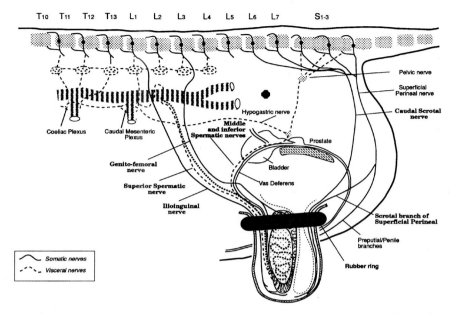

Fig. 6.1 Somatic and visceral nerves of the testes and scrotum affected by application of a rubber ring for castration.

Fig. 6.2 Location of a rubber ring for castration of a lamb of less than 1 week old.

ISCHAEMIC TAIL DOCKING WITH TIGHT RUBBER RINGS

For tail docking the same sized rubber rings are applied with an elastrator to occlude the arterial blood supply and venous drainage of part of the tail. The point of application must leave sufficient tail to cover the anus of male lambs and the vulva of female lambs (Fig. 6.3).

ISCHAEMIC CASTRATION WITH THE BURDIZZO CLAMP

The Burdizzo clamp applies high pressure to the spermatic cord over a narrow (3 to 5 mm) band through the intact skin. The skin, although compressed, is not broken at the time and it usually recovers without its continuity breaking down later. Details of the forces applied and the application time required to ensure irreversible occlusion of the blood vessels in the spermatic cord for different sized lambs are not available. It is common practice to apply the clamp twice to each spermatic cord at separate sites, the second distal to the first, which suggests that there is uncertainty about a single application. A certain amount of skill and awareness of the underlying anatomy are required for effective application of this method and it is not generally used

Fig. 6.3 Application of rubber ring for tail docking of a lamb of less than 1 week old.

for small lambs due to the difficulty of applying the clamp to the spermatic cords with the necessary precision.

In applying the clamp some ischaemia of the scrotum is produced but careful application permits collateral circulation to remain intact.

ISCHAEMIC TAIL DOCKING WITH A BURDIZZO CLAMP

Tail docking using a Burdizzo clamp is not commonly practised in the UK. The Burdizzo clamp may, however, be used to reduce haemorrhage when the tail is docked surgically as described above.

PROBLEMS OCCURRING AT THE TIME OF CASTRATION AND TAIL DOCKING (SHORT TERM)

HANDLING

In order to castrate and tail dock lambs they must be caught and, unless they are already penned, this will require the gathering and restraining of ewes and their lambs. All of this activity interferes, to a greater or lesser extent, with their welfare according to the quality of the methods used, their application and the time involved. These problems of welfare will not be discussed here but it is clear that any cost–benefit analysis of castration and tail docking must include the welfare costs of gathering and handling of the ewes and lambs.

HAEMORRHAGE

The welfare of any animal can be compromised if it suffers a significant decrease in circulating blood volume. The threat to the animal's welfare depends on its state of health at the time, the amount of blood lost and the survival pressures to which it is subjected during and after the haemorrhage. Most haemorrhages occur during or within 24 h of surgical castration. They can be exacerbated by increases in blood pressure and, whenever possible, increased blood pressure as a result of factors such as fear, excitement or heavy exercise, should be avoided. Most methods of surgical castration include attempts to prevent haemorrhage but significant haemorrhage still occurs in some lambs such as those with deficiencies in their haemostatic mechanisms. Visible evidence of haemorrhage is not always present because the spermatic arteries are often broken off within the abdomen when the spermatic cords are twisted and torn.

HERNIATION

If the inguinal canal is open and wider than normal, herniation of abdominal contents can occur and the problem may be exaggerated by increased abdominal pressure, as occurs when the animal struggles or is vigorously restrained. The herniated intestine, mesentery or omentum may be trapped by a rubber ring, crushed by the Burdizzo clamp or released to the outside by surgical incision – all of which can be serious life-threatening problems. These problems can be avoided only by detection of the condition before castration and all animals should be carefully examined before proceeding.

ACUTE PAIN

All the methods of castration and tail docking described above cause acute pain when carried out on unanaesthetized lambs. The intensity and duration of the pain experienced from each method and by different ages of lambs is unclear and will be determined only when methods for quantitative assessment of pain are established. Recent studies by the authors and colleagues using measurements of physiological and behavioural changes related to pain have attempted to establish such methods of assessment and to use them to investigate the three methods described above when applied to lambs of 1, 3 and 6 weeks of age. From these studies and the work of others (Shutt *et al.*, 1988; Mellor and Murray, 1989) it is clear that castration and tail docking with tight rubber rings can provoke changes in behaviour and plasma cortisol which are consistent with the proposition that lambs suffer intense pain in the first hour and that this pain then decreases over the following 1–2 h. The cause of the pain, after the initial nociceptive barrage produced by manipulation of the scrotum, tail, testes and by application of the rubber rings, appears to be a gradually increasing and then decreasing amount of nociceptor activity from the ischaemic tissues. This activity apparently gains access to the central nervous system via intact nerves passing through the rubber rings. The nociceptive barrage increases as the testicular receptors are sensitized and stimulated by substances released into their microenvironment. It has been shown that these nociceptors can function for more than 3 h after their blood supply is occluded (Grubb *et al.*, 1990).

The acute pain from the Burdizzo method appears to be less intense and of shorter duration than that from the rubber ring method. It is presumed that application of the clamp produces an intense but transient barrage of activity both from the afferent nerves in the spermatic cord and from the nociceptors in the scrotal tissues crushed by the clamp. This intense barrage is thought to last for a short time and to subside rapidly since the clamp is expected to exert sufficient pressure to irreversibly damage the afferent nerves and nociceptors and it is not clear how rapidly activity is generated in intact nociceptors by inflammatory responses which follow application of the clamp. No direct recordings are available to confirm these proposals. Some lambs treated with the Burdizzo clamp show signs suggestive of considerable pain which lasts for a relatively long period compared with others treated in what was expected to be the same way, and it appears likely that, in such cases, some afferent nerves continue to conduct.

Acute pain from surgical castration and tail docking can also be of shorter duration and of a different type from that produced by the rubber ring method because, among other things, the testes and part of the tail are removed and cannot provide a continuing source of pain. The acute pain which occurs, is likely to be due to the stimulation of nociceptors and afferent nerves when making the incisions and preventing haemorrhage. A particularly intense barrage appears to be produced by twisting and pulling of the spermatic cord. Again, no direct recordings are available to confirm these proposals. Most of the intense barrage appears to be limited to the time taken to remove the testes and tail. It then appears to be followed, during the ensuing minutes and hours, by increases in activity by intact nociceptors due to inflammation, initially as a result of the trauma and subsequently by the action of pathogens. The changes in behaviour in response to this acute pain are different from those resulting from the application of rubber rings. Lambs reduce their activity and adopt postures which appear to be designed to minimize pain by reducing and avoiding mechanical stimulation of the inflamed and hyperalgesic tissues (Figs 6.4–6.6). This highlights a problem in the assessment of pain, namely that different types of pain elicit different responses.

Under well-controlled conditions, all young lambs respond to a particular method of castration and tail docking in qualitatively similar ways but individual lambs show quantitative differences in their responses. These individual differences are thought to reflect genetic differences and/or differences in the state of development of the underlying physiological mechanisms rather than differences in prior experience of pain particularly in lambs of less than one week of age. No substantial evidence for this is, however, available. In some lambs, psychological effects produced by factors, such as isolation, frightening sensory stimulation or disorientation just prior to castration and

Fig. 6.4 Lamb showing a typical abnormal standing posture soon after castration and tail docking with rubber rings.

Fig. 6.5 Lamb showing severe form of abnormal lying behaviour which included rolling and kicking, soon after castration and tail docking with rubber rings.

tail docking, may lead to blocking of the access of nociceptive activity to the central perceptual mechanisms with the result that little or no pain is experienced. Absence of pain under such circumstances should not, however, be taken to mean that the lamb's welfare is improved because the deleterious effects of these other factors may outweigh the benefits.

REDUCED DISEASE RESISTANCE

All of the acute problems discussed above may lead to neuro-endocrine and immunological changes which reduce the lamb's ability to resist the challenge of pathogenic microorganisms. Only one aspect of this will be mentioned here and that is a particular problem with passive immunity which can occur if castration and tail docking is carried out just after birth. The intake of colostrum may be interfered with since the lamb's behaviour may be dominated by pain and sucking can be prevented for several hours.

PROBLEMS WHICH OCCUR AFTER CASTRATION AND TAIL DOCKING (LONG TERM)

Problems which are secondary to the acts of castration and tail docking can take hours, days or longer to develop. Although

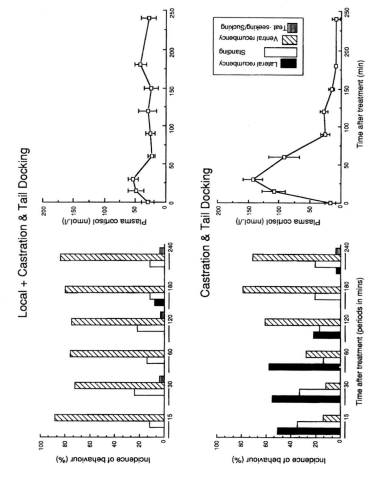

Fig. 6.6 Effect of local anaesthesia of the testes, tail and scrotum on plasma cortisol and behaviour after castration and tail docking with rubber rings. The responses of two groups of six lambs kept inside different pens with their dams are shown. The key indicates the different behaviours recorded for each time period. Note that the lateral recumbency produced by castration and tail docking is almost eliminated by prior treatment with local anaesthetic. The behaviour of the lambs treated with local anaesthetic was not different from a control group of lambs that were just handled (not shown).

many farmers and others are aware of such problems in lambs there have been few systematic studies.

SEPSIS AND INFLAMMATION

The incidence and severity of local infections depend on the ability of particular pathogenic microorganisms to gain access through cuts and other breaks in the skin and on the ability of the lamb to resist such pathogens. If hygiene is good and the lambs are healthy and are kept in clean, dry conditions after castration and tail docking, the incidence and severity of local infections can be low. The problems are made worse if organisms gain access to ischaemic tissue, exudates or blood clots which provide good media for growth and multiplication. The surgical method provides the easiest and most immediate access for pathogens. Accumulation of blood or other fluids in the scrotum can provide a focus for growth and tracking up the spermatic cord can permit access to the peritoneal cavity. Chronic infection of the cut end of the spermatic cord can lead to the condition of scirrhous cord. The Burdizzo method should not normally permit access of pathogenic microorganisms directly from outside but if the skin continuity is broken either at the time of application or if it breaks down later due to ischaemic necrosis, pathogenic microorganisms could gain access to the ischaemic tissues. However, in the absence of pathogenic microorganisms sterile inflammatory processes can occur. The rubber ring method does not normally permit access of pathogens at the time of application but the scrotal skin eventually breaks down and pathogens can gain access to living tissue through the rubber ring.

Depending on the organisms involved, the animal can develop one or more of the following: a mild local infection with limited inflammation and little hyperalgesia, a massive local sepsis and inflammatory reaction, peritonitis, or a systemic infection which may lead to death. Inflammation is usually accompanied by hyperalgesia and by continuing pain but how much these factors affect the well-being of lambs after the various methods of castration is unclear. Intense continuing (chronic) pain in even a small proportion of lambs after castration and tail docking must receive consideration when evaluating the welfare effects of the various methods.

NEUROPATHIC PAIN

Nerve damage can lead to the development of neuropathic pain (akin to phantom limb pain in human amputees) and, under experimental conditions, particular types of damage have produced signs of neuropathic pain in almost 100% of the animals treated (Bennett and Xie, 1988). Because several somatic and visceral nerves are damaged in various ways by the different methods of castration and tail docking it will be necessary to determine if any of these treatments lead to the development of neuropathic pain, and substantial evidence for such problems should be sought.

ANAESTHESIA FOR CASTRATION AND TAIL DOCKING

General anaesthesia is rarely, if ever, used in practice for castration of lambs. However, local anaesthesia can be an effective method for eliminating the acute pain produced by all methods of castration and tail docking provided that all the nerves involved are exposed to an appropriate local anaesthetic for a sufficient time.

If local anaesthetic is infiltrated into the intact testis it is rapidly transported to the spermatic veins and lymphatics from which it can rapidly gain access to the spermatic nerves. If, however, the local anaesthetic is administered after the blood vessels are occluded, for example after applying a rubber ring, then the anaesthetic is retained within the scrotum and distribution occurs by diffusion. Under these conditions it may take a considerable time to block all the afferents from the testis. It is for this reason in particular that local anaesthetic should be used before application of the rubber ring and not after. Occlusion of the blood vessels by the rubber ring keeps the local anaesthetic in the testes and scrotum thus maintaining anaesthesia until after the afferents cease to function. After surgical castration and/or tail docking the local anaesthetic action will be dissipated over its usual time course. Because of the possibility of prolonged pain from some methods it is advisable to use a local anaesthetic with as long an action as possible for these treatments.

Local anaesthesia for tail docking can be rapidly and easily obtained by epidural administration at the coccygeal (C1–2) intervertebral space and the amount administered can be easily controlled to avoid paraparesis.

If local anaesthesia can effectively eliminate the acute pain of castration and tail docking for all of the methods in general use, why is it not used for all such operations? The main reason appears to be economic, i.e. the time and expertise required are considered to make such practice prohibitively expensive.

REGULATION OF CASTRATION AND TAIL DOCKING

In order to improve the welfare of lambs, it has been made an offence to tail dock or castrate lambs which have reached the age of 3 months, without the use of an anaesthetic. Furthermore, the use of a rubber ring or other device to restrict the flow of blood to the tail or scrotum is permitted without an anaesthetic only if the device is applied during the first week of life (Protection of Animals Acts 1911–1988 and, in Scotland, the Protection of Animals (Scotland) Acts 1912–1988). It is obvious that such regulations fail to promote the use of anaesthesia for lambs less than 3 months old, an age which was apparently chosen to accommodate common husbandry practice. There is little quantitative evidence that lambs of less than 3 months suffer sufficiently less than those older than 3 months. While castration and tail docking without anaesthesia by application of tight rubber rings in the first week of life may be supported to some extent by the studies of Barrowman *et al.* (1953, 1954) there are more recent studies, notably those of Shutt *et al.* (1988), Mellor and Murray (1989) and Wood *et al.* (1991), which show that this method can cause pain sufficient to dominate the experience of the lamb for more than 1 h.

CONCLUSION

With increasing awareness and concern for the welfare of animals the practices of castration and tail docking of lambs need continuous consideration. If such practices are necessary then a greater understanding of their physiopathological effects in both the short and long term will be required to enable the most humane approach to be adopted.

REFERENCES

Barrowman, J. R., Boaz, T. G. & Towers, K. G. (1953) Castration of lambs: comparison of the rubber ring ligature and crushing techniques. *Empire Journal of Experimental Agriculture* **21**, 193–203.

Barrowman, J. R., Boaz, T. G. & Towers, K. G. (1954) Castration and docking of lambs: use of the rubber ring ligature at different ages. *Empire Journal of Experimental Agriculture* **22**, 189–202.

Bennett, G. J. & Xie, Y. K. (1988) A peripheral mononeuropathy in rats that produces disorders of pain sensation like those seen in man. *Pain* **33**, 87–107.

Grubb, B. D., Molony, V. & Wood, G. N. (1990) Response of afferents in the superior spermatic nerves of rats to occlusion of the testicular artery and vein. *Pain* (Supplement) **5**, Abstract No. 785.

Mellor, D. J. & Murray, L. (1989) Effects of tail docking and castration on behaviour and plasma cortisol concentrations in young lambs. *Research in Veterinary Science* **46**, 387–391.

Shutt, D. A., Fell, L. R., Connell, R. & Bell, A. K. (1988) Stress responses in lambs docked and castrated surgically or by application of rubber rings. *Australian Veterinary Journal* **65**, 5–7.

Wood, G. N., Molony, V., Hodgson, J. C., Mellor, D. J. & Fleetwood-Walker, S. M. (1991) Effects of local anaesthesia and intravenous naloxone on the changes in behaviour and plasma concentrations of cortisol produced by castration and tail docking with tight rubber rings in young lambs. *Research in Veterinary Science* **51**, 193–199.

CHAPTER 7

Problems of Extensive Sheep Farming Systems

AGNES WINTER

INTRODUCTION

Of all the common farmed species in the UK, sheep are probably perceived by the nonfarming public as being the most "naturally" kept, particularly when farmed extensively on vast areas such as the Pennines, the moors of Devon and Cornwall, the Lake District, the Scottish Highlands and the mountains of Wales. Hardy breeds such as the Herdwick, Swaledale, Welsh Mountain and Scottish Blackface have been developed over centuries to survive the harsh conditions commonly encountered during the long winter months (Fig. 7.1). However, such a system of farming has many risks associated with it despite its generally perceived "green" image.

EXTENSIFICATION SYSTEMS

Flocks which are truly extensively kept throughout the year face a variety of problems such as adverse weather conditions (sudden snowfalls with drifting in winter, wind and rain or snow around lambing time), lack of grazing during the winter,

Fig. 7.1 Winter on Dartmoor – animals in poor condition at the beginning of winter will have difficulty in surviving the harsh conditions (picture supplied by A. Lewis).

and predators. Added to this are the physical difficulties of supplying fodder or concentrates and in adequately inspecting the sheep. Good advance planning is necessary to help reduce the impact of problems which can be foreseen, although the sudden onset of severe weather can still take people (including weather forecasters) by surprise.

The popularity of winter housing (see Fig. 7.2), particularly encouraged when grants were available to assist in building sheep sheds, means that some flocks which are run in a truly extensive system in the summer months may be housed and

Fig. 7.2 Sheep shed typical of many erected on hill farms in the past 15 years.

kept under intensive conditions during the worst months of winter, often until lambing is completed. Thus, the problems associated with intensively housed sheep may be encountered in those same flocks.

"Extensification" may now also simply imply keeping fewer sheep on previously heavily grazed good land, with reduced use of fertilizers and other chemicals. This system should work to the benefit of the animals, particularly as far as internal parasites are concerned. Alternatively, extensification is being promoted on hills to reduce grazing pressure where overgrazing has occurred to the detriment of the environment. However, as well as requiring an overall reduction in sheep numbers, such schemes generally require a further reduction in numbers during the winter months, which results in a concentration of animals on inbye (enclosed) land, increasing stocking rates and, hence, the risk of infectious disease outbreaks. Those farms which have too little inbye land, send ewes off the farm for away-wintering. Unless these animals are well supervised, they may return home in poor condition and bring disease into what are otherwise closed flocks.

FINANCIAL VIABILITY

A major concern with flocks kept in less favoured areas, such as hills and mountains, which are eligible for price support additional to the Sheep Annual Premium (SAP), is the degree to which farm income is dependent upon price support. In 1995, nearly 70% (£28.75 out of £42.41) of the gross margin per ewe was made up of the combined SAP and Hill Land Compensatory Allowance (HLCA) payments (Meat and Livestock Commission 1996). In the most poorly producing flocks, these payments made up almost the whole of the gross margin. There is little doubt that many of those farming such hill and mountain areas would be unable to make a living if price support was significantly reduced or totally withdrawn, unless an alternative support system was introduced.

Only the HLCA payment is dependent upon reasonable standards of husbandry and welfare being practised on the farm.

GENERAL MANAGEMENT OF FLOCKS

INSPECTION OF ANIMALS

The welfare codes for all farmed species, other than sheep, require at least a daily inspection of animals. Although the ways in which traditional extensive systems in some parts of the country are carried out means that sheep are seen regularly (for example, that known as "raking" in Northumberland where sheep are moved from high to lower ground daily), the reduction of manpower on most farms and the physical impossibility of reaching sheep which may have access to thousands of hectares of remote hill or mountain land means that the majority of extensively kept sheep are only seen at gathering; this may happen as few as three times yearly. In such circumstances, the sheep may be regarded almost as feral animals. Provided that the flock is protected against diseases of particular welfare concern, such as scab and blowfly, and is adequately supervised at lambing time, it may be necessary to accept that it is just not possible to see all animals on a regular basis. The increasing use of all-terrain vehicles has, however, made the job of the shepherd easier on all types of farms.

Attempts have been made to quantify the number of ewes for which one person can reasonably be expected to care. Although it is extremely difficult to generalize (since it depends on breed, lambing pattern, availability of housing, etc.), a recent report of the Farm Animal Welfare Council (1994) has suggested that one person should not be responsible for more than about 1000 ewes, with extra labour being required at busy times of the year such as lambing.

NUTRITION

Correct nutrition is of critical importance in extensively managed flocks, particularly at tupping and in the approach to winter. The influence of body condition on ovulation rate is well known (Meat and Livestock Commission, 1981) and, on farms where twin lambs are a definite disadvantage, a narrow path must be trodden between an adequate body condition to support the ewe and one fetus through the worst of the winter weather and the slightly improved body condition which would lead to significant numbers of twin births. The widespread uptake of scanning for fetal numbers now means that twin-bearing ewes can be separated for preferential management. The use of scanning represents one of the most significant advances in the management and welfare of hill ewes in recent years.

Supplementary feeding during the winter months with either fodder, blocks or concentrate feeds can lead to the problem of ewes congregating in the feeding areas, causing damage to the local environment. There may also be problems in the siting of big bale silage feeders, which are required to be well away from water courses to reduce the risk of water contamination. The provision of feedblocks at a number of different well-scattered sites may help to overcome such problems.

Chronic undernutrition caused by overgrazing land in the summer, with ewes entering the winter period in poor condition, yet still left to fend for themselves on the hill, has been a long-standing problem in some areas.

AGE STRUCTURE OF FLOCKS/FITNESS OF EWES

Traditionally, a young age structure has usually been maintained in hill and mountain flocks, with ewes being drafted to less harsh conditions after three to five pregnancies. The early 1990s saw some breakdown in this system, as farmers increased numbers in anticipation of quotas, which were eventually introduced in 1993. Since then, the age structure has largely stabilized, particularly with the eligibility of ewe lambs for the Sheep Annual Premium. However, there is still concern that a minority of farmers have exploited the system by keeping, or even buying in, older ewes and "farming for the subsidy"; furthermore, it is feared that this problem might increase now that unlimited quotas can be leased or purchased. This is additional to the well-established practice of running draft ewe flocks, where older sheep – usually with high lambing percentages, producing crossbred lambs and concentrated at relatively high stocking rates – compound many of the problems associated with extensive sheep farming systems.

Inevitably, keeping older ewes increases the likelihood of dental problems occurring. Incisor loss (broken mouth) is easily recognized, but molar tooth disease often goes unnoticed (Fig. 7.3). Any inadequacy in dentition will have an adverse effect on the ability of ewes to maintain their body condition over the winter months, particularly where they are expected to gain most of their nutrients from grazing.

BREED STRUCTURE

Changes in the breed structure of extensively farmed flocks have taken place or have been contemplated in response to demands for "improved" carcase conformation of slaughter lambs. Where there is selection within traditional breeds for "improved" conformation or decreased fatness, it is important to ensure that these changes do not result in the decreased hardiness of ewes or lambs. Alternatively, the replacement of traditional breeds may be contemplated. This may give rise to similar concerns over reduced hardiness and, furthermore, may precipitate an increased risk of diseases, especially those associated with ticks if unacclimatized sheep are introduced to areas of rough grazing. It also interferes with the traditional "hefting" system, whereby sheep largely keep to the area in which they themselves were reared, ensuring that they are familiar with good grazing or shelter within their home terrain. If changes in breed structure are considered necessary, the method carrying the least risk is to introduce them via the ram line, rather than replace the females with new stock.

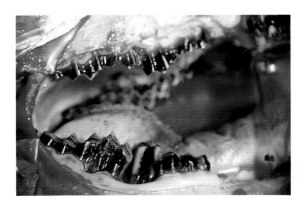

Fig. 7.3 Molar tooth disease often goes unrecognized, but is an important cause of thinness in older sheep.

DISEASE RISKS

Sheep kept extensively the whole year round are likely to be at less risk of some infectious diseases than are intensively kept flocks; in particular, roundworms are unlikely to be a significant problem. Liver fluke may, however, be a major problem on poorly drained land (except on peaty soils where snails are unable to survive). The improvement of hill land by liming and fertilizer application has produced ideal conditions for snails and fluke to flourish. In addition, such pasture improvement has led to deficiencies in trace elements, especially copper, as a result of altering the availability of various mineral constituents of the soil (Fig. 7.4).

Clostridial diseases are also a hazard in extensively managed sheep, with vaccination being a wise precaution. Reports of vaccination reactions affecting, in particular, flocks gathered from extensive grazing conditions, have been investigated. These have largely been attributed to metabolic disease resulting from the stress of rounding up and handling heavily pregnant sheep

Fig. 7.4 Unimproved heather hill (background) and improved land (foreground). The rushy parts of the improved ground provide ideal liver fluke habitat. Copper deficiency may also occur on improved hill land.

combined with their removal from grazing for several hours or longer, or from the effect of administering several treatments such as vaccines, anthelmintics and mineral supplements at the same time (Logue *et al.*, 1989; Wells, 1989). The recommended time of vaccination has been changed from 2 to 6 weeks pre-lambing to try to overcome such problems.

The major disease risks to extensively kept sheep are those caused by external parasites, particularly sheep scab, blowflies, headfly and ticks. Recent concern about the safety to humans of organophosphorus dip compounds has complicated the control of both scab and blowflies, since at present there is no other single product available which gives the comprehensive protection conferred by organophosphorus dips, if correctly used. The situation is particularly worrying in the case of sheep scab, as it takes the presence of only a small number of infected sheep to lead to extensive outbreaks on common grazing, as have already occurred in north and mid-Wales, for example. Millions of sheep are at risk of blowfly strike on all but the highest mountain land.

Alternatives to organophosphorus dips are available, such as flumethrin for scab control, and cypermethrin and cyromazine for blowfly control. The last two compounds are applied as sprays rather than dips. They are effective, but need to be accurately applied and may leave animals only partially protected if not applied over a sufficiently large body area.

In areas of high afforestation, headfly is still a major concern, affecting more animals than blowfly strike. The development of synthetic pyrethroid products which are applied to the head has improved control, but the potential still exists for severe damage to occur if the flies become active before control measures are applied, or if preventive treatments are not repeated throughout the whole of the risk period.

Ticks remain a major problem on large areas of rough grazing, more for the diseases they transmit than for their direct effects on the animals. Small volume applications of synthetic pyrethroids have improved tick control (Fig. 7.5), but lambs can still be badly affected by "cripples" (tick pyaemia) if bitten before protection against ticks is given. It is well recognized that rams and ewes may suffer temporary infertility if first infected with tick-borne fever during the tupping period, and the same disease was thought to be responsible for over 90% of a flock,

Fig. 7.5 The application of synthetic pyrethroid products has improved tick control.

recently introduced onto tick-infested land, aborting (Jones and Davies, 1995).

LAMBING AND LAMB MANAGEMENT

LAMB MORTALITY

Lamb mortality in flocks lambing out of doors is highly dependent upon weather conditions around lambing, with losses of up to 45% having been reported (Slee, 1976). The spring of 1994 was notable for prolonged periods of wet, cold weather and there were many reports in the farming press of exceptionally high mortality in both indoor and outdoor lambing flocks. Outdoor losses are largely caused by hypothermia, whereas losses indoors are caused by infections, spread being exacerbated by the need to keep ewes and lambs in for longer, often leading to overcrowding. The practice in some hill flocks of lambing ewes on limited inbye land may, in some circumstances, lead to losses from both categories of disease: infections because of the close confinement of animals on land which may become contaminated in bad weather, and hypothermia because lambs are still exposed to adverse weather conditions. The provision of simple shelter can do much to reduce losses associated with hypothermia and does not require extra labour as the work can be carried out before lambing begins.

The frequency of inspection of ewes lambing in extensive conditions may pose problems. Many farmers feel that available labour is best targeted as those animals which are most likely to require help, such as maiden ewes and those identified on scans as carrying twins. Although there is obviously a slight risk in less frequent (once daily only, say) inspection of mature native breeds which are likely to produce a single lamb with the minimum of effort, in practice few problems may be encountered. The risks, of course, will be that much greater if scanning has not been carried out, the ewes have been overfed (resulting in excessively large lambs) or rams likely to produce lambs with large heads have been used.

TIMING OF ROUTINE PROCEDURES

One major managemental problem for flocks lambing in extensive conditions is the timing of routine procedures such as castration and tailing. Current regulations limit the use of the rubber ring to lambs under 1 week old. Where flocks lamb in extensive conditions, the tendency is not to gather the sheep until lambs are at least 3 weeks old because of the risks of injury and mismothering. Fear of prosecution has led to some farmers reverting to open castration with the knife (Fig. 7.6); when performed by the experienced shepherd, this may pose few risks, but it may potentially cause losses from haemorrhage, infections and the occasional intestinal prolapse. An improved bloodless castrator also offers an alternative method.

There is ongoing debate about methods of castration, the possibility of provision of analgesia and, indeed, over the necessity to castrate at all (Farm Animal Welfare Council, 1994; see also Chapter 6). Some farmers have found that they do not need to castrate lambs if they can be sold at low weights and a significant trade to other EU countries has built up. However, problems with live "exports" means that the future of this outlet for light male hill lambs may be in some doubt.

Fig. 7.6 Open castration may be carried out on lambs above the age limit for the use of rubber rings.

DEALING WITH CASUALTIES

Even on the best run sheep farms, there will inevitably be casualties, particularly around lambing time. On extensive farms where veterinary help, an abattoir willing to deal with casualties, or knackers may be many miles away, it is particularly important that someone takes responsibility for these animals, either giving appropriate first aid, or preventing further suffering by being willing and able to kill both adult animals and lambs humanely. Shooting is the recommended method for adult sheep, while a sharp blow to the back of the head is effective for lambs.

ENVIRONMENTAL CONSIDERATIONS

In summary, extensive sheep farming has many potential problems for both the animals and farmers. It is highly dependent on price support and on weather conditions. If price support was to be reduced or withdrawn, it is likely that many hill and mountain farms would face severe financial hardship. A major reduction in sheep numbers would have a significant effect on the environment which at present is largely sheep created. Such change might not necessarily be for the better, since there is evidence that merely removing sheep from hill land does not guarantee an improvement in habitat, but can result in an increase in undesirable species of vegetation.

ACKNOWLEDGEMENTS

The author thanks Professor M. Clarkson and B. Merrell for helpful comments.

REFERENCES AND FURTHER READING

Farm Animal Welfare Council (1994) *Report on the Welfare of Sheep*, FAWC, Tolworth, Surbiton.

Jones, G. L. & Davies, I. H. (1995) An ovine abortion storm caused by infection with *Cytocetes phagocytophila*. *Veterinary Record* **137**, 127.

Logue, D. N., Brodie, T. & Bogan, J. A. (1989) Downer ewe syndrome associated with prelambing dosing of heavily pregnant ewes. *Proceedings of the Sheep Veterinary Society* **14**, 89–93.

MAFF (1990) *Code of Recommendations for the Welfare of Sheep*. MAFF Publications, London.

Meat and Livestock Commission (1981) *Feeding the Ewe*. MLC, Milton Keynes.

Meat and Livestock Commission (1996) *Sheep Yearbook*. MLC, Milton Keynes.

Sheep Veterinary Society (1994) *The Casualty Sheep*. (Available from SVS Secretariat). Moredun Institute, Edinburgh.

Slee, J. (1976) Cold stress and perinatal mortality in lambs. In *Veterinary Annual*, 16th edn, pp. 66–69. Wright, Bristol.

Wells, P. W. (1989) Suspect adverse reactions following vaccination of pregnant ewes. *Proceedings of the Sheep Veterinary Society* **14**, 85–88.

CHAPTER 8

Anaesthesia in Sheep and Goats

POLLY TAYLOR

INTRODUCTION

Sheep and goats are the poor relations where general anaesthesia is concerned. However, equipment used for small animals is suitable for these species and, with attention to the special requirements of ruminants, sheep and goats can be anaesthetized successfully. Most of the commonly used anaesthetic and sedative drugs do not carry a product licence for sheep and, especially, goats. However, those described in the text have been tested in clinical and experimental settings and all the recommendations made are based on wide experience of many authors; their use is entirely justified on welfare grounds.

SPECIAL CONSIDERATIONS FOR RUMINANTS

All the general principles that apply to any animal that is to be given an anaesthetic apply to ruminants. Most general anaesthetics, and to a lesser extent the sedatives, induce respiratory and cardiovascular depression and remove normal protective reflexes such as coughing and temperature control. It is essential at all times to ensure a clear airway (**A**, airway), adequate ventilation (**B**, breathing) and adequate circulation (**C**, circulation). There are some additional problems in ruminants that are not encountered in simple-stomached animals which require consideration when anaesthesia of the sheep or goat is contemplated (see Table 8.1).

REGURGITATION

The rumen cannot be emptied by pre-anaesthetic starvation and large volumes of rumen content may be regurgitated during anaesthesia. Regurgitation is always a possibility during sheep and goat anaesthesia and precautions must be taken to ensure that rumen contents are not aspirated. Pre-anaesthetic starvation for 18–24 h (or overnight, 12–18 h) should reduce the likelihood of regurgitation by reducing the volume of rumen content. Starvation for very much longer than this may result in a higher liquid component and make regurgitation more likely. Endotracheal intubation with a cuffed tube should be regarded as essential whenever general anaesthesia is induced in an adult ruminant. This is a good general rule, although it may not be strictly necessary for very short procedures and sometimes when ketamine is used.

RUMINAL TYMPANY

Whenever normal erructation is prevented by anaesthesia and recumbency, gas will build up in the rumen inducing bloat. This

Table 8.1 General anaesthesia: problems in ruminants.

Regurgitation	Always intubate the trachea. Starve 12–18 h
Salivation	Tilt head. Collect saliva. Replace with equal volume of Ringer's lactate and bicarbonate
Bloat	Starve 12–18 h. Stomach tube/paracentesis. Sternal recumbency
Hypoxaemia/hypercapnia	Lateral recumbency. At least 30% inspired oxygen. IPPV
Recovery	Sternal recumbency. Supervision

is a serious problem under general anaesthesia as ruminal distension reduces lung capacity and impairs ventilation. Because respiration is already likely to be compromised by general anaesthesia bloat may be fatal. Pre-anaesthetic starvation reduces the amount of gas produced by fermentation and is usually all that is required to prevent any serious ruminal distension, especially important for animals that have been on lush pasture. Long periods of anaesthesia may necessitate release of ruminal gas even after adequate starvation. Passing a stomach tube may sometimes release sufficient gas, but abdominal paracentesis is required if the rumen becomes obviously distended and the stomach tube is unproductive.

SALIVATION

Ruminants salivate continuously and this does not stop during general anaesthesia (Fig. 8.1). Ruminant salivation is little affected by anticholinergic agents unless very high doses are used. In this case the saliva becomes thick and mucoid and is more difficult to deal with. Copious salivation must be regarded as an immovable feature of ruminant anaesthesia and must be prevented from entering the trachea. This reinforces the need to intubate whenever general anaesthesia is performed. It is beneficial to support the head at the poll in animals in lateral recumbency, so that saliva runs out of the mouth and does not collect in the pharynx. Loss of large volumes of alkalotic fluid will induce hypovolaemia and acidosis. In practice, this is only a real hazard during very long procedures (more than 2–3 hours). In this case the saliva should be collected and replaced

Fig. 8.1 A ruminant continuing to salivate throughout anaesthesia.

on a volume for volume basis with lactated Ringer's solution and bicarbonate, 1 mmol/kg.h (N.B. 4.2% bicarbonate contains 0.5 mmol/ml).

CARDIOVASCULAR AND RESPIRATORY EMBARRASSMENT

In lateral and, more particularly, dorsal recumbency the large mass of rumen and intestines may cause respiratory and cardiovascular embarrassment due to pressure on the diaphragm and great vessels. Cardiovascular effects can be minimized by keeping any periods of dorsal recumbency as short as possible, as it is primarily in this position that serious compression of the posterior vena cava and abdominal aorta occurs. Pressure on the diaphragm is also most serious in dorsal recumbency, but still occurs in lateral recumbency. Pressure on the diaphragm leads to decreased tidal volume and inadequate ventilation and also increased areas of lung tissue that are not ventilated. These lead to respiratory acidosis and hypoxaemia. Hypoxaemia can be prevented by the provision of at least 30% inspired oxygen and should be provided in all but the very short (less than 10–15 min) procedures. Respiratory acidosis will be compounded by the respiratory depressant effects of most anaesthetic drugs and is best prevented by the use of intermittent positive pressure ventilation (IPPV). IPPV is essential when long periods of anaesthesia (more than 1–2 h) are required. In general, lower inspired concentrations of any inhaled anaesthetics are required during IPPV.

RECOVERY

The same hazards of regurgitation, salivation, aspiration and ruminal distension are present during the recovery period until the animal has regained good control of pharyngeal and laryngeal reflexes and can sit up. A patent airway must be maintained during this period and this is best done by leaving the endotracheal tube in place until strong swallowing reflexes return. The animal should be supported in sternal recumbency until it can maintain its own position. This will both reduce the chance of aspiration while the animal is still weak and release ruminal gas. The animal must be supervised in this period.

ANALGESIA

Sheep and goats are seriously neglected when it comes to clinical analgesia and there are no analgesics licensed for them (see Table 8.2), nor are withdrawal times available. However, there is no reason to believe that, unlike other species, sheep and goats do not require analgesia for painful surgical conditions. Sheep have been used in experimental pain studies and buprenorphine (6 to 10 µg/kg), butorphanol (0.2 mg/kg) and the α-2 agonists (xylazine 50–100 µg/kg, detomidine 10–20 µg/kg and clonidine 6 µg/kg) appear to be effective in a number of situations. The opiates are most effective for visceral pain and last several hours. The α-2 agonists are effective in all types of pain but do not last as long. The non-steroidal anti-inflammatory drugs (NSAIDs) are not licensed for use in sheep or goats, but limited clinical use indicates that flunixin or carprofen 2 mg/kg daily provides good pain relief following orthopaedic surgery and can be supplemented with opioids such as buprenorphine. The NSAIDS flunixin, carprofen and ketoprofen are licensed for use in cattle and can legally be used in other animals intended for human consumption according to the "Cascade" system (RCVS, 1996). Some form of postoperative analgesia should be provided at the very least for orthopaedic surgery. Where there is concern about using nonlicensed drugs, regional analgesia is a good alternative. There is much evidence

Table 8.2 Analgesics (no product licences).

Group	Drug	Dose	Duration of effect	Use
Opioids	Butorphanol	0.2 mg/kg	Lasts 2–3 h	Visceral pain
	Buprenorphine	10 µg/kg	Slow onset, lasts 4 h	Visceral pain
α-2 agonists	Xylazine	50–100 µg/kg	Lasts 45 min	All pain
	Detomidine	10–20 µg/kg	Lasts 60 min	All pain
NSAIDs	Flunixin	2 mg/kg	Lasts 12–24 h	All pain
	Carprofen	2 mg/kg	Lasts up to 24 h	All pain

indicating that a regional block applied before surgery improves the postoperative analgesia. Local analgesia can be safely used in conjunction with general anaesthesia.

SEDATION

A number of procedures can be performed in sheep and goats under regional analgesia with sedation. Sedatives are also used for premedication (Table 8.3).

AGENTS TO USE

Acepromazine (0.05–0.1 mg/kg) provides mild sedation and is often sufficient to prevent a naturally phlegmatic animal from becoming restless during a long procedure. It does not provide analgesia. It causes hypotension in the hypovolaemic animal and is best avoided in such cases. It smooths anaesthetic induction and recovery with barbiturates but substantially increases the likelihood of regurgitation.

Xylazine (sheep 100–200 µg/kg; goats up to 50 µg/kg) induces deep sedation. Up to twice as much may be used if given intramuscularly. Goats appear to be more sensitive to the drug than sheep and there is some breed variation, particularly in sheep. The smaller, more primitive breeds appear to require the higher dose rates. Xylazine induces bradycardia which lasts throughout sedation, although it is most marked immediately after intravenous injection. Xylazine induces marked hypoxae-

Table 8.3 Sedatives.

Drug	Dose	Effect
Acepromazine	0.05–0.1 mg/kg	Mild sedation Increased risk of regurgitation
Xylazine	100–200 µg/kg (sheep) Up to 50 µg/kg (goats)	Profound sedation Hypoxaemia
Diazepam	0.25–0.5 mg/kg	Moderate sedation

mia in sheep and probably also in goats. It should not be used in animals with respiratory disease, and supplementary oxygen should be given (e.g. by mask) if long procedures are performed with the sheep in lateral recumbency. Xylazine is successfully used as premedication before ketamine (as described below). α-2 antagonists (xylazine is an α-2 agonist) have been used successfully in sheep and goats, and atipamezole (Antisedan, Pfizer) is now available for use in dogs and cats in the UK. It does not have a product licence for sheep and goats but is extremely effective at reversing the effects of xylazine in these species.

Diazepam (0.25–0.5 mg/kg) has been found useful in many clinics, particularly in goats. It is given slowly intravenously and provides 30 min of good sedation. The animal can be quite easily roused and there is no analgesia.

Anticholinergic agents are not sedatives but may be used with sedatives or as premedication. Their effect on salivation has been described and their use for routine premedication is controversial. Doses of 0.4 mg/kg can be used to block vagal reflexes which may be induced by traction on abdominal viscera or pressure on the eye. Atropine is also preferred by some when xylazine is used in order to prevent bradycardia.

GENERAL ANAESTHESIA

INTUBATION

Intubation is most easily performed under direct vision with a laryngoscope (Fig. 8.2). A large blade for human or small animal use is adequate. The easiest method is with the animal in sternal recumbency with the head and neck fully extended. Alternatively, with the animal supported in dorsal recumbency (Fig. 8.3), an assistant holds the tongue out of the way and pulls the upper jaw down. The laryngoscope is inserted and lifted until the epiglottis falls forward exposing the larynx. The tube is then passed between the cords. Sternal recumbency is preferable for cardiovascular stability and makes control of saliva or rumen content during intubation easier. Intubation can also be performed without the aid of a laryngoscope and the animal in lateral or dorsal recumbency. The larynx is grasped from the

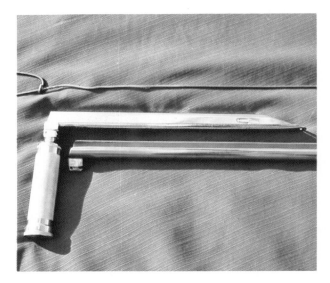

Fig. 8.2 Laryngoscope and tube stiffener (metal stilette shown here, but stiff plastic also good). The upper laryngoscope is suitable for most goats and sheep. The longer blade (Roswon laryngoscope) is designed for cattle but may occasionally be needed for very large goats.

outside and "threaded over" the tube as it is passed into the back of the mouth. This technique is the most difficult, but useful when a laryngoscope is not available. All techniques are easier if anaesthesia is deep enough for the animal to be relaxed and not chewing. A wire or rigid plastic stilette passed inside the tube to within a few centimetres of the tip stiffens the tube and makes it easier to pass through the larynx (Fig. 8.2). Young goats are prone to potential fatal laryngospasm and the larynx should be sprayed with local anaesthetic solution at least 30 s before intubation is attempted.

INTRAVENOUS AGENTS

Thiopentone is a short acting barbiturate best given without premedication at 15 mg/kg as a bolus. The trachea should be intubated immediately. If used without premedication regurgitation does not commonly occur during intubation. Acepromazine premedication may allow a slightly lower dose of thiopentone to be used. Apnoea lasting 30–60 s is common after thiopentone and a few assisted breaths may be required. This dose of thiopentone provides only a few minutes anaesthesia

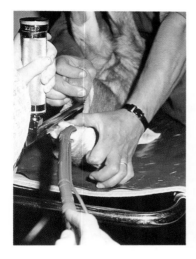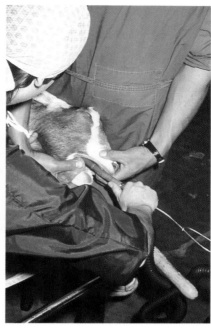

Fig. 8.3 (Left) Intubation using a laryngoscope, with the animal in dorsal recumbency. (Right) "Blind" intubation with the animal in lateral recumbency.

and is suitable for induction before inhalation anaesthesia (see Table 8.4).

Pentobarbitone is metabolized more rapidly in sheep than in other domestic species and, if used at 20 mg/kg, induction is similar to that with thiopentone. It is best given more slowly than thiopentone. The trachea should be intubated immediately, and conditions are similar to those with thiopentone. Recovery from one dose of pentobarbitone is as rapid as after thiopentone and it can be used in the same way for induction before inhalation anaesthesia. Post induction apnoea is slightly less marked than with thiopentone unless the injection was too rapid. Anaesthesia can be maintained for short periods with incremental pentobarbitone (approximately 2 mg/kg every 5 min). The drug is cumulative, however, and recovery will be prolonged if anaesthesia is maintained in this way for long periods (more than 30–40 min). Some respiratory depression is likely and hypoxaemia should be prevented with supplementary oxygen

Table 8.4 Intravenous agents.

Drug	Dose	Effect
Thiopentone	15 mg/kg	Induction or very short procedure
Pentobarbitone	20 mg/kg	Induction or short procedure Incremental doses prolong recovery
Methohexitone	4 mg/kg	Very short-acting
Ketamine		Best in combination with xylazine or diazepam (see text)
Alphadolone acetate and alphaxalone (Saffan)	3 mg/kg	Induction and maintenance. Good for incremental doses.
Propofol (Rapinovet)	4–6 mg/kg	Induction and maintenance. OK for incremental doses.

via the endotracheal tube. Commercially available solutions of pentobarbitone contain propylene glycol, reputed to cause haemolysis in sheep and goats. Solutions should be made up from powder using water or saline and up to 10% ethanol.

Methohexitone is an ultra short-acting barbiturate which may be used in preference to thiopentone because of its shorter recovery. Induction with 4 mg/kg methohexitone is similar to induction with thiopentone, but there is less time for intubation. Anaesthesia can be prolonged with further doses of 50–75 mg/min but recovery from methohexitone alone is associated with twitching or convulsive activity unless sedative premedication has been given. Respiratory depression is similar to that seen with the other barbiturates.

Ketamine has proved popular for sheep and goat anaesthesia and can be used on its own (10–15 mg/kg, slowly intravenously), but high muscle tone and trembling makes the effect unpleasant, although analgesia appears to be good. Better results are achieved when ketamine is combined with xylazine or diazepam. Doses of xylazine 200 μg/kg intramuscularly followed after 15 min by 10 mg/kg ketamine slowly intravenously produce 40–45 min of anaesthesia. This involves quite a high dose of xylazine for goats and 0.1 mg/kg xylazine intramuscularly followed after 10 min with 5 mg/kg ketamine intravenously (slowly) may be preferable. This gives 15–20 min of anaesthesia. Intubation of the trachea is not essential when this technique is used and oxygen can be given by face mask. However, intubation as a precaution against aspiration can easily be performed under xylazine–ketamine anaesthesia and is advisable at least for abdominal surgery where handling the viscera may cause regurgitation or passive reflux. Diazepam 0.5 mg/kg intravenously followed by 4 mg/kg ketamine intravenously results in a shorter period of anaesthesia and is particularly useful for induction before gaseous anaesthesia because of minimal respiratory depression. Xylazine (50 µg/kg intravenously or 100 µg/kg intramuscularly) followed by ketamine (5 mg/kg intravenously) is also a good technique for induction and intubation before inhalation anaesthesia. To reduce the amount of xylazine given, half may be replaced by 0.2 mg/kg diazepam before induction of anaesthesia with 5 mg/kg ketamine.

Alphadolone acetate and alphaxalone (Saffan, Mallinckrodt) have proved to be safe anaesthetics in sheep and goats both for induction and maintenance of anaesthesia. Doses of 3 mg/kg provide good conditions for intubation and give about 10 min anaesthesia. Some respiratory and cardiovascular depression is seen at induction but it is relatively transient and less than with the barbiturates. Saffan has proved useful both for induction prior to inhalation anaesthesia and for maintenance of anaesthesia with infusions of 2–2.5 mg/kg.min. In both instances recovery is smooth and rapid.

Propofol (Rapinovet, Mallinckrodt) is commercially available for dogs and cats and is proving useful for induction of anaesthesia in sheep and goats (4–6 mg/kg slowly, intravenously). It is probably preferable to thiopentone if cost does not preclude its use.

INHALATION ANAESTHESIA

Once the trachea has been intubated maintenance of anaesthesia with a volatile agent is similar to that in other species. Adult goats and sheep can be anaesthetized using adult human nonrebreathing circuits such as the Magill (Fig. 8.4) or co-axial circuits. Rebreathing systems are also suitable, in particular circle systems. It is difficult to position a sheep or goat on a standard operating table with a Waters canister (which should be as close to the animal's mouth as possible to keep dead space

Fig. 8.4 Inhalation anaesthesia using a Magill circuit.

to a minimum) and soda-lime absorption is more efficient in a circle system.

Halothane is the most commonly used inhalation agent in sheep and goats. It is used both for induction of anaesthesia using a mask (Fig. 8.5) and, more commonly, for maintenance of anaesthesia after intravenous induction. It causes cardio-vascular and respiratory depression as in other species. For long procedures (over 1–2 h) IPPV may be necessary to prevent severe respiratory acidosis. Intravenous fluids can be used to support the circulation. Other volatile agents have been used in sheep and goats and the faster recovery seen after isoflurane makes it a suitable agent, if cost allows.

Nitrous oxide can be used in sheep and goats but is best avoided in rebreathing circuits unless high flow rates are used or the inspired oxygen concentration can be measured, so that dangerously low concentrations of oxygen are not allowed to develop. Nitrous oxide may also increase ruminal tympany.

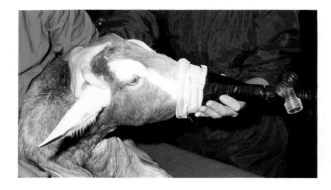

Fig. 8.5 Induction of anaesthesia by inhalation of halothane using a mask.

MAINTENANCE AND MONITORING

Monitoring of anaesthesia should be performed as in any other species. Depth of anaesthesia is assessed by attention to vital signs and the response to surgery. The eye may rotate ventrally but this is not a consistent sign and the more classical effects of anaesthesia on the cardiovascular and respiratory systems are a more reliable guide. Temperature should be monitored in long procedures and in newborn animals. More specialized techniques for measuring blood pressure, electrocardiography, and tidal carbon dioxide and blood gases can be used to provide additional information in the same way as in other species. The middle auricular artery is particularly accessible for direct arterial blood pressure monitoring. Fluid therapy to replace existing deficits, provide daily requirements and make up losses occurring during anaesthesia should be provided as for any other species.

LOCAL ANAESTHESIA

Sheep and goats are ideally suited to local anaesthetic techniques under sedation or manual restraint. Many procedures can be carried out in this way, and there is a wealth of experience available from both clinical and research backgrounds. Lignocaine is the most commonly used of the local anaesthetic solutions and is well tolerated in both species. Some of the newer solutions that cause even less tissue reaction may be used in sheep and goats but offer no particular advantage in these species.

Lignocaine is toxic at high doses and convulsions occur at around 6 mg/kg intravenously or 10 mg/kg intramuscularly. Convulsions are usually preceded by drowsiness and respiratory depression. The total dose given by local infiltration should be kept below 10 mg/kg. A 1 or 2% solution is suitable for most adult sheep and goats.

Local infiltration can be used in sheep and goats as in other animals. The most useful regional blocks for these species are the cornual block, L-block, and paravertebral block, caudal epidural and intravenous regional anaesthesia of the limb (see Figs 8.6–8.9 and Table 8.5).

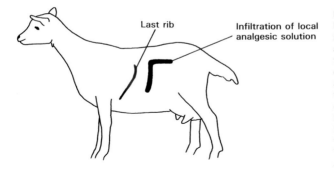

Fig. 8.6 The L-block. Local anaesthetic solution is infiltrated into the skin and the full thickness of the body wall in the configuration shown. The area behind and below the inverted L is desensitized, making this block suitable for flank laparotomy. Care must be taken to avoid overdose. The abdominal contents are not affected and the viscera must be handled carefully to avoid tension on the mesentery or intestinal walls.

Whenever local anaesthesia is employed strict asepsis must be maintained, particularly for epidural block. Adrenaline 1:100,000 to 1:200,000 is a useful addition to the lignocaine solution as it delays absorption, both prolonging the block and reducing the likelihood of toxicity. Adrenaline should not be included when the solution is to be used for ring block of small structures (e.g. teats) as it may induce ischaemia.

SPECIAL CONSIDERATIONS FOR LAMBS AND KIDS

Anaesthesia is commonly required in the very young for horn disbudding. A number of features of the neonatal kid and lamb must be taken into account. As in any species the neonatal lamb and goat have little metabolic reserve and are susceptible to starvation and cold. The time away from the dam should be kept to a minimum and pre-anaesthetic starvation is best kept to an hour or two or avoided altogether. Heavy sedatives are best replaced with short-acting anaesthesia so that the animal returns to normal activity as soon as possible. It must not be allowed to get cold, and good insulation (e.g. Flectabed lamb jackets) or heating pads should be used in any procedure needing more than a very few minutes of general anaesthesia.

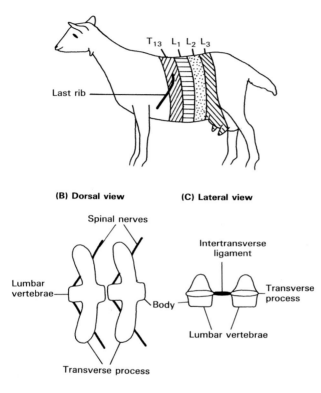

(A) Paravertebral block
to show area served by each of T_{13}-L_3

T_{13} L_1 L_2 L_3

Last rib

(B) Dorsal view   **(C) Lateral view**

Spinal nerves

Lumbar vertebrae

Body

Transverse process

Intertransverse ligament

Transverse process

Lumbar vertebrae

Fig. 8.7 The paravertebral block. Suitable for flank laparotomy. Some abdominal viscera affected. (A) shows the area served by each of the 13th thoracic or first three lumbar nerves. (B) shows the nerves leaving the spinal cord in the intervertebral space anterior to the vertebra of the same segment. They run caudally at an angle and are best blocked by "walking" the needle over the anterior edge of the transverse process approximately half way between the midline and the tip of the transverse process. The skin and intervening muscle is infiltrated with local anaesthetic solution and then 3–5 ml of 1 to 2% lignocaine is injected once the needle tip has passed through the intertransverse ligament (see C).

SEDATIVES

The sedatives already described act in a similar manner in the young as in adults, with the notable exception of xylazine, where the dose should be reduced by half for lambs or to 25 μg/kg intramuscularly in goats. Diazepam (0.1–0.2 mg/kg) is an extremely effective and safer alternative in the newborn.

GENERAL ANAESTHESIA

Barbiturates are best avoided in the neonatal lamb and kid as they do not metabolize them as rapidly as adults, particularly pentobarbitone.

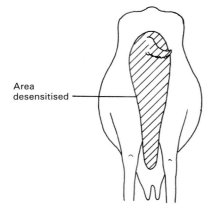

Area
desensitised ————

Fig. 8.8 Caudal epidural. The shaded area is desensitized when a caudal epidural block is used. Strict asepsis is required. Between 1 and 4 ml (according to the size of animal) of 2% lignocaine are injected into the 1st coccygeal space. The needle is held at right angles to the skin and a loss of resistance to injection occurs when the needle enters the epidural space. Lumbar epidural is induced by injection of larger volumes in the lumbosacral space (see Hall and Clarke, 1983).

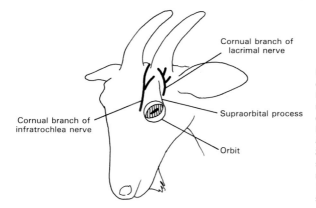

Cornual branch of
lacrimal nerve

Cornual branch of
infratrochlea nerve

Supraorbital process

Orbit

Fig. 8.9 Nerve block for dehorning. Both the cornual branch of the infratrochlea nerve and the cornual branch of the lacrimal nerve must be blocked for dehorning goats and disbudding kids. Two to 3 ml of 2% lignocaine at each site for the adult goat and not more than 2 ml (preferably 1 ml) of 0.5% solution at each site in young kids.

Saffan (2–6 mg/kg) has proved ideal for disbudding very young goats. One induction dose is usually sufficient for the whole procedure but can be topped up with smaller doses if necessary. Recovery is complete and rapid and the animal can usually be returned to its dam within a few minutes of completing the procedure. Xylazine–ketamine (with reduced doses of xylazine) or diazepam–xylazine have also been used but the recovery is not quite so rapid. Anaesthesia is particularly slow after xylazine and ketamine unless the xylazine can be reversed with an antagonist such as atipamezole.

Inhalation induction of anaesthesia with halothane has proved well tolerated in neonatal lambs and kids. Anaesthesia can be continued using a mask or after intubation. If a mask is

Table 8.5 Intravenous regional anaesthesia of the foot (see Edwards, 1981).

- Good for lower limb surgery
- Sheep or goat in lateral recumbency
- Apply tourniquet above hock or carpus
- Inject 2–5 ml 2% lignocaine (no adrenaline) into lateral saphenous, lateral plantar or lateral plantar digital vein (any vein distal to tourniquet that can be located)
- Anaesthesia occurs within 10 min
- Normal sensation returns within a few minutes of removal of tourniquet: keep tourniquet on throughout surgery
- Potential toxicity: do not remove tourniquet for at least 15–20 min after injection of local anaesthetic solution

used throughout anaesthesia for disbudding it must be removed when a gas-burning hot iron is used for disbudding or the animal's face may be set alight. Despite this problem inhalation anaesthesia has proved an excellent technique for disbudding neonatal goats. Recovery is smooth, rapid and complete, so time away from the dam is minimal.

LOCAL ANAESTHESIA

Small lambs, and kids in particular, have a reputation for extreme sensitivity to local anaesthetic solutions. It is probably a feature of their size and they have simply been overdosed. Severe toxicity (convulsions) occurs at approximately 10 mg/kg lignocaine intramuscularly, equivalent to 2 ml of 2% lignocaine in a 4 kg kid. If a local analgesic block is to be used for disbudding newborn kids it is best to dilute the lignocaine (see Fig. 8.9) to 0.5% so that 1–2 ml can still be used at each site.

ACKNOWLEDGEMENTS

To Dr Avril Waterman for helpful advice during preparation of the manuscript. Thanks to the Goat Veterinary Society for pictures.

REFERENCES AND FURTHER READING

Buttle, H., Mowlem, A. & Mews, A. (1986) Disbudding and dehorning of goats. *In Practice* **8**, 63–65.

Edwards, G. B. (1981) Intravenous regional anaesthesia of the bovine foot. *In Practice* **3**, 13–14.

Gray, P. R. & McDonell, W.N. (1986) Anaesthesia in goats and sheep. Part I Local analgesia. *Compendium on Continuing Education for the Practicing Veterinarian* **8**, 33–39.

Gray, P.R. & McDonell, W.N. (1986) Anaesthesia in goats and sheep. Part II General analgesia. *Compendium on Continuing Education for the Practicing Veterinarian* **8**, 127–134.

Hall, L. W. & Clarke, K. W. (1983) *Veterinary Anaesthesia.* pp. 273–286. Baillière Tindall, London.

Kumar, A., Thurmon, J. C. & Hardenbrook, H. J. (1976) *Veterinary Medicine/Small Animal Clinician* **271**, 1707–1713.

Riebold, T. W. (1996) Ruminants. In *Lumb and Jones' Veterinary Anaesthesia*, 3rd edn. (eds, Thurman, J.C., Tranquilli, W.J. & Benson, G.J.), pp. 610–626. Williams and Wilkins, Baltimore.

Royal College of Veterinary Surgeons (1996) RCVS Guide to Professional Conduct 1996. pp. 62–64. RCVS, London.

Waterman, A. E. (1981) Evaluation of the actions and use of alphaxolone/alphadolone (CT1341) in sheep. *Research in Veterinary Science* **30**, 114–119.

Waterman, A. E. (1988) Use of propofol in sheep. *Veterinary Record* **122**, 260.

CHAPTER 9

Parasite Control in Sheep

GERALD COLES

INTRODUCTION

The UK has the largest and most successful sheep industry in the European Union and produces a product (lamb) that is exported to other countries within the EU (Fig. 9.1). Parasites pose a major threat to sheep health and productivity. They fall into three quite distinct groups, the protozoa (such as coccidia

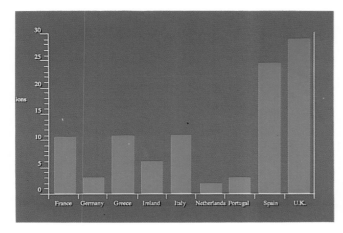

Fig. 9.1 Relative sizes of the national sheep flocks within the European Union.

and *Toxoplasma*), the helminths (tapeworms, fluke and nematodes) and the ectoparasites (mites, keds, lice and flies).

Parasite control in sheep is achieved largely by the use of chemicals (see Fig. 9.2). Caution is required, however, as haphazard use of chemicals poses the risk of development of resistant strains of parasites.

PROTOZOA

COCCIDIA

Coccidia are host specific. Eleven species have been reported in sheep, but only two, *Eimeria crandallis* and *E. ovinoidalis*, are considered important pathogens. The oocysts are highly resistant and can remain infective on the ground for long periods of time. The parasites have an enormous reproductive potential and the degree of infection is related to the number of oocysts ingested. Potentially, one oocyst could destroy all the villous epithelial cells in 2 m of gut. Damage is more serious in the large intestine where crypt cells may be destroyed.

Coccidiosis, which appears to be of growing concern to farmers as a result of increasing intensification and indoor lambing, can be incorrectly diagnosed. Oocysts in the faeces must be sporulated and identified to check if high numbers of pathogenic species are present – just seeing oocysts does not mean

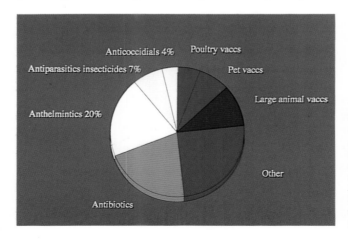

Fig. 9.2 The importance of chemicals to the UK sheep industry is reflected in antiparasitics making up the single biggest sector of the animal health market. (Figure excludes feed additives.)

coccidial disease. In heavy infections, serious damage to the gut and disease can occur before oocysts are shed.

Control

The best method of control is good hygiene. This includes keeping areas around the feeding trough clean and dry and providing adequate clean bedding for housed sheep. Young lambs grazed shortly after others have contaminated the pasture are most at risk.

Coccidiosis can be controlled by the inclusion of anticoccidials (decoquinate, Deccox; Rhône Mérieux) in the diet of ewes and/or lambs to achieve approximately 1 mg/kg.day for 28 days. It can also be controlled by inclusion of anticoccidials in water (e.g. amprolium; unlicensed, by veterinary prescription only). These regimens are not very effective, however, as voluntary intake has two problems: first, young lambs do not reliably take creep feed or water; second, if lambs are kept coccidia-free they do not develop immunity and can therefore develop coccidiosis when medicated feed is withdrawn.

Infected animals are treated with amprolium; in severe outbreaks the dose may be doubled. Sulphonamides can also be used by injection (see Chapter 10). Because damage to the gut has often occurred by the time of diagnosis and treatment, recovery may not be rapid and may be sufficiently delayed to increase days to slaughter and the proportion of "tail-end" lambs.

CRYPTOSPORIDIA

Cryptosporidiosis is caused by a very small coccidium with a direct life cycle. It results in scouring and loss of appetite and can be a particular problem in orphan lambs. The infection is a zoonosis and there have been a number of cases of children becoming infected from handling lambs, especially on farms open to the public. Good hygiene and washing of hands is therefore important. Infection occurs by ingestion of oocysts and is most common in very young lambs. It can prevent lambs feeding for 3 or 4 days. While lambs housed indoors can recover from this set back, for those outside in cold or wet conditions the temporary cessation of feeding may be fatal.

Control

Currently, no effective chemotherapy is available, but decontamination of infected housing can be undertaken using an ammonia-based disinfectant (e.g. Cocide, Antec International).

TOXOPLASMA

Toxoplasma gondii is an important cause of abortion in ewes. Infection in early pregnancy can result in fetal resorption, while primary infection of the ewe from days 50–102 often causes premature birth of stillborn or weakly lambs. Once ewes are infected they remain so, but transfer of the parasite to the fetus happens only when a primary infection occurs during pregnancy. Infection of ewes is usually a result of ingestion of feed contaminated by oocysts excreted in the faeces of young infected cats. Diagnosis in the sheep is by inspection of the placental cotyledons which have characteristic small white foci of necrosis; this, combined with the history of the flock, may indicate toxoplasmosis. Laboratory confirmation is often required.

Control

Once sheep have been infected they are subsequently immune and should therefore be kept. If animals become infected 3 weeks prior to pregnancy no problems should arise, so allowing animals to become naturally infected is an option. Protection can now be provided by vaccination with a live attenuated strain of *Toxoplasma* (Toxovax, Mycofarm) which can be given from 5 months of age or, in older animals, in the 4 months prior to mating. This is preferable to the inclusion of monensin at 15 mg/kg during pregnancy – monensin is a toxic compound and doses must not be exceeded. A more promising treatment is a combination of sulphamethazine and pyrimethamine, but this is not yet licensed for use in sheep.

HELMINTHS

TAPEWORMS

The adult stage of *Moniezia expansa* (less commonly *M. benedini*) infects lambs. Because the segments are visible in faeces they cause concern to the farmer. But, with the possible exceptions of intestinal blockage (unproven) and increasing the risk of blowfly strike through softening the faeces, published evidence that *Moniezia* causes problems in lambs in the UK is not available.

Control

Control is with modern benzimidazoles (see Table 9.1). Treatment initiated when segments are seen in the faeces will remove most or all of the worms.

LIVER FLUKE

Heavy infection by *Fasciola hepatica* is usually only seen in the autumn (acute fluke disease) and is diagnosed by post mortem examination (Fig. 9.3). Lighter infections (chronic fluke disease) are diagnosed by the presence of fluke eggs in the faeces.

Control

The disease can be controlled by drainage and fencing off of wet areas. The spraying of snail habitats with molluscicides is also effective but is not now widely practised. Control is more usually achieved by the use of fasciolicides. However, even regular use of fasciolicides will not entirely eradicate infections, as rabbits act as a natural reservoir of infection.

A range of drugs control adult fluke (see Table 9.2). A number are sold as combination products or are benzimidazoles with fasciolicidal activity and are effective against nematodes as well as fluke. Because not all fluke may be mature at any one time,

Table 9.1 Benzimidazoles for the control of tapeworm in sheep.

Drug	Dose rate (mg/kg)	Trade name
Albendazole	5	Albex (Chanelle) Endospec (Bimeda) Tramazole (Tulivin) Valbazen (Pfizer)
Febantel	5	Bayverm Armadose (Bayer)
Fenbendazole	5	Fendazole (Bimeda) Fenzol (Norbrook) Panacur (Hoechst) Wormaway (Diversey Lever) Zerofen (Chanelle)
Netobimin	7.5	Hapadex (Schering-Plough)
Mebendazole	15	Chanazole (Chanelle) Ovitelmin (Janssen Animal Health)
Oxfendazole	5	Bovex (Chanelle) Oxfenil (Virbac) Parafend (Norbrook) Systamex (Mallinckrodt)
Ricobendazole	5	Allverm (Crown) Bental (C-Vet) Rycoben (Young's Animal Health)

those fasciolicides with activity against younger fluke are advantageous.

Resistance to closantel has been reported in fluke in England and Wales and triclabendazole resistance has been found in fluke in sheep and cattle in Ireland. Apparent reduced efficacy of fasciolicides in sheep should be investigated with quantitative fluke egg counts.

NEMATODES

Nematodes remain the parasites of most concern to sheep farmers. Infections are usually mixed, with *Ostertagia* (*Teladorsagia*) *circumcincta* being the most important species. The relative importance of the different species may vary with the time of year. *Nematodirus battus* causes problems in young lambs in the

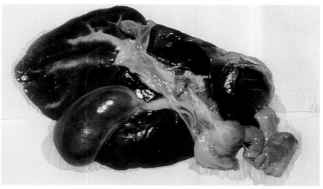

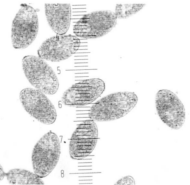

Fig. 9.3 (Top) Chronic fluke infection of the liver showing enlarged bile ducts. (Left) Eggs of the liver fluke *Fasciola hepatica*.

spring. *Haemonchus contortus* is the most pathogenic nematode (Fig. 9.4), causing anaemia and, with heavy infections, death. The free-living stages prefer warm moist weather – *H. contortus* is, therefore, most common from July to September and is mainly confined to the southern counties of the UK. *Trichostron-gylus vitrinus* can cause disease in the autumn, particularly in ewes and hogg replacements.

The nematode life cycle is direct, eggs hatching with adequate warmth and moisture, but eggs can accumulate in dry weather leading to heavy infections after rain. The young larvae feed on bacteria in the faeces and moult to give infective third-stage larvae. Infective larvae can survive for many months on pasture but overwintering larvae usually die out by early June. Important exceptions are *N. battus* which usually overwinters as eggs and *H. contortus* which usually overwinters in sheep as inhibited larvae. Ingested larvae normally develop into adults

Table 9.2 Spectrum of activity of fasciolicides for sheep.

Drug	Dose rate (mg/kg)	Trade name	Effective against			
			Flukes		Nematodes	
			Adult	Immature	Normal	Benzimidazole resistant
Benzimidazoles						
Albendazole	10	Albex (Chanelle)				
		Endospec (Bimeda)	X		X	
		Tramazole (Tulivin)				
		Valbazen (Pfizer)	X		X	
Netobimin	20	Hapadex (Schering-Plough)	X		X	
Ricobendazole	7.5	Allverm (Crown)	X		X	
		Bental (C-Vet)	X		X	
		Rycoben (Young's Animal Health)	X		X	
Triclabendazole*	10	Combinex+ (Novartis)	X	X	X	X
		Fasinex (Novartis)	X	X		

Salicylanilides and nitrophenols							
Closantel	10	Flukiver (Janssen)		X	X		
		Flukol (Young's)		X	X		
		Supaverm[+] (Janssen)				X	
Nitroxynil	10	Trodax (Rhône Mérieux)		X	X	X	
	15						
Oxyclozanide	15	Klomisole[+] (Bimeda)	X	X		X	X
		Levafas[+] (Norbrook)	X	X		X	X
		Nilzan[+] (Mallinckrodt)	X	X		X	X
		Systamex Plus Fluke[+] (Mallinckrodt)	X	X		X	
		Wormaway Levamisole	X				
		Oxyclozanide Drench[+] (Diversey Lever)	X			X	X
	45	Zanil (Mallinckrodt)	X				
		Zanil (Mallinckrodt)	X	X	X		

× Active against.
[*] Although chemically a benzimidazole, triclabendazole has a different mechanism of action to benzimidazoles.
[+] Combination product (flukicide plus anthelmintic).

Fig. 9.4 Adult *Haemonchus contortus* in the abomasum.

in about 3 weeks; larvae which are inhibited in the autumn resume development in the spring or at the time of parturition (periparturient rise).

Control

Certain uncoupling agents (closantel, nitroxynil) can be used for the control of *H. contortus* but not for other species of nematodes. Control measures largely rely on the administration of anthelmintics. There are three groups of broad spectrum anthelmintics each with a different mechanism of action (see Table 9.3).

Group 1 (1-BZ), the benzimidazoles and probenzimidazoles (which cyclize to benzimidazoles in the host), act by affecting tubulin polymerization. Only group 1 anthelmintics are ovicidal (kill nematode eggs).

Group 2 (2-LM), levamisole and morantel, act on acetylcholine receptors.

Group 3 (3-AV), ivermectin and moxidectin, and shortly likely to include doramectin, are thought to affect chloride ion movement in the γ-aminobutyric acid receptors, but the exact mechanism of action is not known.

Resistance in nematodes to any one anthelmintic in a group means resistance to all drugs in that group (e.g. all

Table 9.3 Classes of anthelmintics for the control of nematodes in sheep.

Drug	Dose rate (mg/kg)	Trade name
Class 1: Benzimidazoles and probenzimidazoles		
Albendazole	5	Albex (Chanelle)
		Endospec (Bimeda)
		Tramazole (Tulivin)
		Valbazen (Pfizer)
Febantel	5	Bayverm (Bayer)
Fenbendazole	5	Fendazole (Bimeda)
		Fenzol (Norbrook)
		Panacur (Hoechst Roussel)
		Wormaway (Diversey Lever)
		Zerofen (Chanelle)
Mebendazole	15	Chanazole (Chanelle)
		Ovitelmin (Janssen)
		Supaverm* (Janssen)
Netobimin	7.5	Hapadex (Schering-Plough)
Oxfendazole	5	Bovex (Chanelle)
		Oxfenil (Virbac)
		Parafend (Norbrook)
		Systamex (Mallinckrodt)
		Systamex Plus Fluke (Mallinckrodt)
Ricobendazole	5	Allverm (Crown)
		Bental (C-Vet)
		Rycoben (Young's Animal Health)
Thiophanate	In feed	Molamin Worming Bucket (Dallas Keith)
		Nemafax (Rhône Mérieux)
		Wormalic (Dallas Keith)
		Provitblock (Dallas Keith)
Class 2: Levamisole and morantel		
Levamisole	7.5	Armadose Breakwormer (Bayer)
		Bionem (Mallinckrodt)
		Chanaverm (Chanelle)
		Combinex (Novartis)
		Decazole (Bimeda)
		Klomisole (Bimeda)
		Levacide (Norbrook)
		Levacur (Hoechst)
		Levadin (Vetoquinol)
		Levafas* (Norbrook)
		Nilverm (Mallinckrodt)
		Nilzan* (Mallinckrodt)
		Ridaverm (Crown)
		Ripercol (Janssen)
		Sure (Young's Animal Health)

Continued.

Table 9.3 Continued.

Drug	Dose rate (mg/kg)	Trade name
		Vermisole (Bimeda)
		Wormaway (Diversey Lever)
Morantel	11.6	Exhelm (Pfizer)
Class 3: Avermectins and milbemycins		
Ivermectin	0.2	Ivomec injection (MSD Agvet)
		Oramec (MSD Agvet)
		Panomec (MSD Agvet)
		Rycomec (Young's Animal Health)
Moxidectin	0.2	Cydectin (Cyanamid)
Doramectin[‡]		

[*] Combination product containing a fasciolicide
[‡] To be registered.

benzimidazoles are side resistant). The unplanned use of anthelmintics, therefore, cannot be recommended and can lead to the development of resistant populations of worms. Control should be based on epidemiological principles (see below).

Controlling the spread of anthelmintic resistance

The control of nematodes is being complicated by the development of anthelmintic resistance. Resistance to class 1 anthelmintics, the benzimidazoles, is widespread, affecting about half the sheep farms in southern England and two-thirds of non-dairy goat farms in England and Wales (see Table 9.4). Benzimidazole-resistant nematodes have also been recorded in Scotland, but the prevalence is less than in England. It has been shown in the Netherlands that on many sheep farms with benzimidazole resistance there are no genes for susceptibility left. This will explain why, once benzimidazole resistance has become a problem, reversion to susceptibility does not occur. The relationship between results of *in vitro* tests for benzimidazole resistance and loss of all genes for benzimidazole susceptibility in a population of worms has not been established.

Levamisole resistance has been confirmed on a Devon and a Buckinghamshire sheep farm and in two Angora goat herds,

Table 9.4 Occurrence of benzimidazole-resistant (BZ-R) nematodes in randomly selected sheep flocks and non-dairy goat herds in England and Wales.

Animal	Date	County	Flocks with BZ-R nematodes (%)
Sheep	1990	East Sussex	35
Sheep	1990	Oxfordshire	44
Sheep	1990	West Sussex	61
Sheep	1992	Devon and Cornwall	44*
Sheep	1992	Northumberland and Durham	15*
Goats	1992	England and Wales	65

Ostertagia circumcincta 69%, *Haemonchus contortus* 27%, *Trichostrongylus* species 2%, *Cooperia curticei* 2%.

but is probably more common than the results suggest. Ivermectin resistance has been reported from an experimental goat herd in Scotland and two Angora goat herds in south-west England. *O. circumcincta* in the Angora goat herds was resistant to all three groups. The adoption of strategies to control the spread of resistance is therefore essential. These have been discussed in detail by Coles and Roush (1992). In summary they are:

(1) The correct dose should be administered: Underdosing permits the survival of partly resistant worms which can interbreed to produce resistant worms. Some animals should be weighed and the dose should be calculated, according to the manufacturer's recommendations, based on the heaviest in the group. The accuracy of the dosing gun should be checked using a plastic measuring cylinder.
(2) The minimum number of treatments should be used: The more often animals are dosed the faster resistance can develop. An Australian perspective is that more than three treatments per year will eventually produce problems of resistance.
(3) The type of anthelmintic should be changed annually (NOT more often): Annual rotation, probably best conducted in the spring, may be the single most important thing a farmer can do to delay the development of resistant nematodes. Despite the many products on the market there are only three different broad spectrum pharmacological classes of anthelmintics available (see Table 9.3). Where benzimidazole resistance is present, alternation should be between group 2, levamisole or morantel,

and group 3, ivermectin or moxidectin. When benzimidazole–
levamisole resistance is found alternation should be between a
benzimidazole plus levamisole and group 3. A commercial mixture may be marketed shortly, but until this is available products should be given sequentially and not mixed. Ivermectin
should not be used continuously year after year as resistance
can develop relatively rapidly (e.g. after 16 doses in goats).

(4)   Resistant nematodes should not be purchased with animals. The movement of animals may be the major cause of the
spread of resistance. Animals should be treated with a group 3
anthelmintic, ivermectin, moxidectin or shortly doramectin, on
arrival, yarded for at least 30 hours, dipped in a scab-approved
dip to prevent introduction of scab and lice, and checked by a
veterinary surgeon for infectious diseases before joining the
main flock. Because benzimidazole resistance is common, benzimidazoles cannot be used for animals being brought on to
the farm.

(5)   Sheep should never be kept on the same paddocks as goats
or follow goats on a grazing rotation: Resistant nematodes are
more common in goats than in sheep.

(6)   Farms should be checked for resistance regularly: The use
of ineffective drugs is a waste of money and will only heighten
the problem of resistance. A faecal egg count reduction test
(FECRT) should be performed 10–14 days after treatment with
an anthelmintic. The reason for the time-lag is that benzimidazoles can stop worms laying eggs without removing the
worms – worms must, therefore, be given time to resume egg
laying. False positives can be obtained with levamisole due to
the failure of levamisole always to remove all immature worms.
If the FECRT suggests levamisole failure, the observation
should be repeated collecting faecal samples four days after
treatment, and/or confirmatory laboratory tests should be run;
these tests – the egg hatch test for benzimidazole resistance and
the larval development test for benzimidazole and levamisole
resistance – can be obtained via veterinary investigation centres
or the University of Bristol "Worm Check" service.

(7)   Epidemiological control of nematodes should be undertaken on farms with the aid of a veterinary surgeon: The four
principal strategies are outlined below.

Epidemiological control of nematodes

Epidemiological control is based on preventing the build up of large numbers of infective larvae on pasture from midsummer onwards derived from the contamination passed by ewes and young lambs in the spring. Four main strategies are available:

(1)   Use of clean pasture: Ewes are dosed with an effective anthelmintic prior to turn out on to pasture not used the previous year for sheep or goats. A single treatment (using a drug which is effective against inhibited larvae) when ewes are taken inside or before turn out should suffice for worm control. However, some species of nematodes (*N. battus, Cooperia* species) can be transmitted by calves, and *Trichostrongylus axei* infects cattle and horses as well as sheep. Most sheep farmers do not have adequate clean pasture/new leys in the spring.

(2)   Treating lambs at the end of June and moving to clean pasture (usually an aftermath): Most sheep farmers will be making hay or cutting silage and, if not grazed since lambing, these fields will be "clean" by the end of June. Lambs treated with nonbenzimidazoles should be held on uninfected pasture or a yard for 30 h before moving to the new field to prevent contamination of the clean pasture with eggs in the intestines. Group 2 and 3 anthelmintics kill the worms but not the eggs.

(3)   Spring suppression: Because it takes about 3 weeks for ingested larvae to develop into egg-laying adults, lambs and ewes can be used to "clean" contaminated pasture by treating them every 3 weeks from turn out. The last dose must be in June to ensure that pasture infectivity has reached a low level (i.e. earlier turn out requires more treatments). Three-weekly doses prevent significant contamination of the pasture and further treatments after June should not be required.

Because of the long persistency of action of moxidectin 3-weekly dosing is not required and as few as two doses may prevent later build-up of pasture contamination, provided that *N. battus* is not a problem. Although moxidectin kills *N. battus*, it does not provide extended protection against reinfection. An alternative system was the use in the ewe of an albendazole bolus, Proftril (Pfizer), but this has been withdrawn from the market. In Australia, when the bolus was used with benzimidazole resistant nematodes, breakdown of nematode control occurred within 3–4 years due to the selection of highly resistant worms. It is possible that an ivermectin bolus will be produced for use in sheep in the UK. But the full benefit of the bolus will not be obtained unless lambs are treated at the same time as the bolus is protecting the ewe.

(4)   Early lambing: Lambs sold before the end of June should not need treating with anthelmintics unless *N. battus* is a problem.

Use of an anthelmintic to which there is already some resistance will effectively select for much higher levels of resistance if strategies 1, 2 or 3 are adopted: the worms contaminating the pasture are those that have survived treatment. Therefore, control strategies should not be initiated before it has been confirmed that the anthelmintic to be used is effective; that is, that there is no detectable resistance on the farm to the product to be used.

ECTOPARASITES (Table 9.5)

BLOWFLY

Blowflies (*Lucilia sericata* and *L. caesar*) (Fig. 9.5) primarily deposit their eggs on detritus and carrion but can lay hundreds on the fleece of a single sheep, often as a response to fleece soiling or fleece rot. Blowfly strike can cause very serious

Fig. 9.5 Blowfly larvae and adult fly (Picture N. French, University of Bristol).

welfare problems: heavy infestations can kill through toxicity within 3 days of oviposition. Although only a small percentage of animals in one flock is usually affected, protection of the whole flock is advisable from late May onwards. A diagnosis is confirmed by finding eggs or maggots of the blowfly in wounds or wet areas of the fleece.

Control

Preventive measures include dipping in organophosphates (diazinon, propetamphos), spraying the relevant areas with cypermethrin or use of a cyromazine pour-on. Two treatments during the fly season should usually suffice, although the length of protection will depend on the wool length and varies between types of treatment, dipping affording longer protection than other treatments. Active lesions can be treated with pyrethroid pour-ons (cypermethrin or deltamethrin) but these give little or no protection as pour-ons. Cypermethrin given as a spray does, however, give 6–8 weeks protection against blowfly strike.

NASAL BOT FLY

Two generations of adult nasal bot fly (*Oestrus ovis*) each year deposit larvae in the nose of the sheep. Autumn infestations overwinter in the ewe, while early summer infestations develop rapidly. Infestation is sporadic but can be severe in some flocks. The most characteristic symptoms are nose rubbing and nasal discharge (symptoms can be confused with scrapie).

Control

Animals showing symptoms can be treated at any time of the year with oral closantel or ivermectin. To eliminate infestations, all animals should be treated between December and February; neighbouring flocks must be treated at the same time to prevent reinfestation.

Table 9.5 Spectrum of activity of ectoparasiticides for sheep.

Drug	Blowfly	Nasal bot fly	Head fly	Sheep scab	Lice	Keds	Ticks
PYRETHROIDS*							
Cypermethrin							
Crovect Pour-On§ (Crown)	X		X	Dip only	X		X
Crovect Dip (Crown)¶							
Cypor+ (Young's Animal Health)							
Provinec Pour-On§ (C–Vet)							
Provinec Dip (C–Vet)¶							
Robust Dip (Young's Animal Health)¶							
Vector Pour-On§ (Young's Animal Health)							
Deltamethrin							
Coopers Spot On+ (Mallinckrodt)	X		X		X	X	X
Flumethrin							
Bayticol Scab & Tick Dip (Bayer)				X	X	X	X
Coopers Green Label Scab & Tick Dip (Mallinckrodt)							
ORGANOPHOSPHATES							
Diazinon	X§			X	X	X	X
Coopers All Season Fly & Scab Dip (Mallinckrodt)							
Diazadip All Seasons (Bayer)							

Product						
Deosan Diazinon Dip (Diversey Lever)						
Gold Fleece (Bimeda)						
Paracide Plus (Battle, Hayward & Bower)						
Summer flydip (Battle, Hayward & Bower)						
Topclip Fly and Scab Dip (Novartis)	X	X	X	X	X	
Propetamphos						
Ectomort Centenary (Young's Animal Health)	X	X	X	X	X	
Flyte 1250 (Young's Animal Health)						
Seraphos Scab Approved Sheep Dip (Crown)	X	X	X			
OTHERS						
Amitraz						
Taktic (Hoechst Roussel)	X					
Closantel						
Flukiver (Janssen)						
Supaverm (Janssen)						
Cyromazine						
Vetrazin Pour-On (Novartis)	X‡					
Ivermectin						
Ivomec injection (MSD Agvet)	X	X				
Oramec Drench (MSD Agvet)	X					
Panomec (MSD Agvet)	X	X				

X Active against.

* Pyrethroids are applied to the strike for blowfly treatment, as a spray for blowfly prevention, and as a pin-stream down the back line for ticks and lice.

¶ Cypermethrin-based dips (but not pour-ons) control sensitive sheep scab. Blowfly: ‡ Treatment only. ‡ Prevention only. § Prevention and treatment.

HEAD FLY

Although present in most areas, head fly (*Hydrotaea irritans*) primarily causes "broken head" lesions in sheep in northern England and southern Scotland and is a particular problem in the horned Scottish Blackface. The injuries are probably self-inflicted by the sheep in attempting to avoid the irritation of the fly. Wounds are strongly attractive to this fly, which feeds on blood, and can attract secondary infections. The head fly, and other blood-feeding species, can be responsible for the spread of diseases such as keratoconjunctivitis, mastitis, infectious vulvo-vaginitis and even border disease.

Control

Cypermethrin pour-on or spray or deltamethrin spot-on, though not 100% effective, are used in the control of head fly. One or two treatments during the head fly season (June/July) should suffice. Many farmers still depend on tar-type repellents which present a physical barrier to injury.

SHEEP SCAB

Sheep scab, a form of sheep mange, is a highly infectious disease causing characteristic lesions and wool loss (Fig. 9.6). Infec-

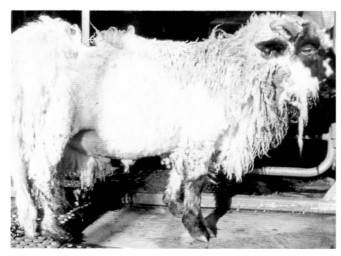

Fig. 9.6 Sheep scab (*Psoroptes ovis*) infestation.

tion can kill untreated animals – the most pathogenic isolates can do so within 4 or 5 weeks and may also cause epileptic seizures. The disease is an allergic dermatitis caused by the excretory products of the mite, *Psoroptes ovis*. Until recently, outbreaks were mostly associated with dealers or common grazing. However, since dipping was deregulated, it was estimated in 1996 (on the basis of pelt damage) that scab has increased from around 1% to 15% of animals infested with peaks up to 25%. Infection should be suspected if there is widespread rubbing, nibbling with wool hanging from the teeth, mouthing or general fleece deterioration. It is confirmed by the presence of mites, but care must be taken not to confuse *P. ovis* with forage mites.

Control

The disease is controlled by dipping in MAFF scab-approved dips containing the organophosphates, diazinon and propetamphos, or the pyrethroid, flumethrin. A high *cis* cypermethrin dip can be used for cure but it does not protect against reinfestation for as long as the MAFF-approved products. There is increasing public concern about the use of organophosphate sheep dips due to alleged poisoning of dip operators and the environmental problems of dip disposal. Aside from the direct signs of inhibition of acetylcholinesterase in humans, organophosphates can cause delayed neuropathy. Because of the importance of proper operator protection, training and certification for the use of sheep dips is now required.

Spray and pour-on treatments are not effective and should not be used. Injection with two doses of ivermectin (200 µg/kg) given seven days apart is effective and is the preferred treatment during the winter and for ewes in lamb. Although ivermectin injection does not protect against reinfestation, moxidectin injection, a product expected to be registered shortly, does provide protection.

PSOROPTIC EAR MITES

Psoroptic ear mites are occasionally found and are likely to be a subpopulation of the scab mite. Infestation with ear mites is not associated with any very characteristic symptoms, although the presence of an excessive wax plug in the affected ears can result in conspicuous head-shaking in some animals. Psoroptic mites found on cotton wool buds used to clean the ear will confirm the diagnosis. It is possible, but unproven, that ear mites may occasionally re-establish on the body of sheep and cause sheep scab, which may explain occasional outbreaks of sheep scab in closed flocks.

Control

Currently, the best method of control of ear mites is by ivermectin injection given subcutaneously at 200 µg/kg. This may only be given under veterinary prescription as the product is not licensed for this purpose in sheep.

LICE AND KEDS

The annual dipping against sheep scab over the past years has had a profound effect on the other ectoparasites of sheep – the body louse (*Bovicola* (formerly *Damalinia*) *ovis*) (Fig. 9.7), blue louse (*Linognathus ovillus*), ked (*Melophagus ovinus*) and chorioptic mange mite (*Chorioptes ovis*) have rarely been seen. Based on pelt damage, lice have increased from 1% of pelts infested to around 5% by 1996 and keds have reappeared.

Wool biting and fleece staining are the primary features of ked infestation, which can have a serious effect on wool pro-

Fig. 9.7 The effects of infestation with the sheep louse *Bovicola ovis.*

duction. Lice also cause loss of condition through rubbing and scratching by animals. Both should be controlled on welfare grounds. It is suspected that organophosphate-resistant strains of lice may have developed in the Pennine area. Diagnosis is by identification of the parasites on the sheep where wool appears to be in poor condition.

Control

Lice and keds can be controlled by the use of organophosphate (diazinon and propetamphos) or pyrethroid (flumethrin) dips. An alternative method of control is to use a pyrethroid pour-on (cypermethrin) or spot-on (deltamethrin). Because currently available pour-ons are not usually 100% effective, there is a strong selection pressure for the development of insecticide resistance. In Australia, widespread resistance of the sheep louse to pyrethroid pour-ons has developed. It is important to ensure that strains resistant to both organophosphates and pyrethroids do not develop in the UK or the control of chewing lice might prove difficult or impossible.

TICKS

Heavy infestations of ticks (primarily *Ixodes ricinus*; rarely *Dermacentor reticulatus* or, in coastal regions, *Haemaphysalis punctata*) can be a problem in their own right in young animals, but disease transmission is probably more important. Three-host ticks, such as *I. ricinus*, feed on a wide range of animals and then drop off their hosts. Each stage lasts 1 year and non-feeding stages require the moist environment of a thick mat of vegetation, as provided by rough unimproved pastures typically found in hill country. Diseases of sheep carried by ticks include louping ill, rickettsial "tick-borne fever" and tick pyaemia (*Staphylococcus aureus*). Ticks also transmit the zoonoses Lyme disease and Q fever.

Control

Control measures include dipping in diazinon, propetamphos or flumethrin, the use of pour-ons (cypermethrin) and spot-ons (deltamethrin), or amitraz.

REFERENCES AND FURTHER READING

Brunsdon, R. V. (1980) Principles of helminth control. *Veterinary Parasitology* **6**, 185–215.

Buxton, D., Thomson, K. M. & Maley, S. (1993) Treatment of ovine toxoplasmosis with a combination of sulphamethazine and pyrimethamine. *Veterinary Record* **132**, 409–411.

Coles, G. C. & Roush, R. T. (1992) Slowing the spread of anthelmintic resistant nematodes of sheep and goats in the United Kingdom. *Veterinary Record* **130**, 505–510.

Gregory, M. W., Catchpole, J., Nolan, A. & Hebert, C. N. (1988) Ovine coccidiosis: studies of *Eimeria ovinoidalis* and *E. crandallis* in conventionally reared lambs, including possible effects of passive immunity. *Deutsche Tierarztliche Wochenschrift* **96**, 287–292.

Hong, C., Hunt, K. R. & Coles, G. C. (1996) Occurrence of anthelmintic resistant nematodes on sheep farms in England and goat farms in England and Wales. *Veterinary Record* **139**, 83–86.

Mathieson, A. O. (1991) Ectoparasites and their control. In *Diseases of Sheep*. 2nd edn. (eds Martin, W. B. & Aitken, I. D.), pp. 284–290. Blackwell Scientific Publications, Oxford.

Tarry, D. W. (1991) Sheep scab and other forms of mange. In *Diseases of Sheep*. 2nd edn. (eds Martin, W. B. & Aitken, I. D.), pp. 261–265. Blackwell Scientific Publications, Oxford.

CHAPTER 10

Diagnosis and Control of Coccidiosis in Sheep

MIKE TAYLOR

INTRODUCTION

Coccidiosis is a frequently diagnosed, but often misunderstood, parasitic infection in sheep. Most sheep are infected with coccidia during their lives but in the majority of animals the parasites cause little or no damage. Disease occurs only if animals are subjected to heavy infections, or if their resistance is lowered. It is, therefore, important to differentiate between infection and disease, as the presence of coccidia does not invariably lead to the development of clinical signs of disease. Indeed, trials have demonstrated that low levels of challenge can be beneficial by stimulating protective immune responses in the host.

PARASITE LIFE CYCLE

Lambs become infected with coccidia through the ingestion of sporulated oocysts (Fig. 10.1). From each oocyst, eight sporozoites emerge in the small intestine and penetrate cells in the intestinal mucosa. The parasites undergo at least one asexual multiplication within the mucosa giving rise to merozoites within

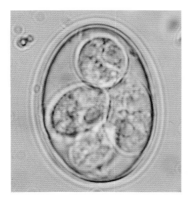

Fig. 10.1 Oocyst of *Eimeria ovinoidalis*, the most pathogenic species of coccidia seen in sheep.

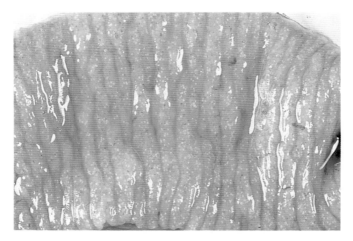

Fig. 10.2 Visible 'white spots' in the small intestinal mucosa are giant first-generation schizonts (meronts) of *E. ovinoidalis*.

schizonts (meronts) (Fig. 10.2). In most species of sheep coccidia, the first-generation schizonts are very large (100–300 μm) and may be visible to the naked eye as pinpoint white spots on the mucosa. These give rise to a second generation of schizonts which are much smaller than the first. From the last schizont generation, merozoites emerge which give rise to the sexual forms (gamonts) which in turn form oocysts that pass out in the faeces. Once outside, the oocysts sporulate, i.e. they undergo two divisions to produce four sporocysts, each containing two sporozoites. Only the sporulated oocysts are infective. If ingested by a susceptible host, the sporozoites emerge and start the cycle again.

PATHOGENICITY

In sheep, some species of coccidia are more pathogenic than others. Coccidia that invade the small intestine generally produce fewer pathogenic effects. This is because in ruminants the small intestine is very long, providing a large number of host cells and allowing the potential for enormous parasite replication with minimal damage. If the absorption of nutrients is impaired, the large intestine is, to some extent, capable of compensating. Coccidia that invade the large intestine are more likely to cause pathological changes, particularly if large numbers of oocysts are ingested over a short period of time. Here, the rate of cellular turnover is much lower and there is no compensatory effect from other regions of the gut. The result is that water resorption is impaired, leading to diarrhoea.

Those coccidia that produce small endogenous stages and develop in epithelial cells of the villi produce minimal lesions. In contrast, coccidia that infect stem cells in the crypt epithelium may produce more extensive lesions by preventing the replacement of damaged epithelium. Coccidia with large endogenous schizogony stages can evoke either localized reactions or more diffuse lesions because of the large numbers of merozoites released into the tissues once the schizonts mature and rupture. This eventually leads to the development of numerous gametocyte stages and oocysts which can be the most damaging because of the sheer numbers present.

Eimeria ovinoidalis is the most pathogenic species of coccidia seen in sheep in the UK. This parasite affects the large intestine of the host, has large endogenous stages and can kill crypt stem cells.

The effects of coccidial infection may be exacerbated if different species that affect different parts of the gut are present at the same time. Similarly, concurrent infections with other disease-producing agents such as helminths, bacteria and viruses may affect the severity of disease; interactions between coccidia and *Nematodirus battus* have been shown to have particularly damaging effects in young lambs.

HOST RESISTANCE

Colostrum provides passive immunity to coccidiosis during the first few weeks of life. Thereafter, susceptibility to *E. ovinoidalis* and *Eimeria crandallis* has been found to increase progressively in lambs up to 4 weeks old. Subsequently animals acquire resistance to coccidia as a result of active immunity.

While sheep of all ages are susceptible to infection, younger lambs are more susceptible to disease. The majority of lambs will probably become infected during the first few months of life and may or may not show signs of disease. Those that reach adulthood are highly resistant to the pathogenic effects of the parasites but may continue to harbour small numbers throughout their lives.

Occasionally, acute coccidiosis occurs in adult animals with impaired cellular immunity of in those which have been subjected to stress, such as dietary changes, prolonged travel, extremes of temperature and weather conditions, changes in environment or severe concurrent infection. An animal's nutritional status, and mineral and vitamin deficiencies can also influence resistance to infection. Suckling animals, in addition to benefiting from colostral intake, may forage less and hence pick up fewer oocysts from pasture. Well-nourished animals may simply be able to fight off infection more readily.

CLINICAL AND PATHOLOGICAL SIGNS

There are 11 species of coccidia that are recognized in sheep in the UK (Table 10.1), most of which can be distinguished by oocyst morphology. The first sign that coccidiosis may be affecting a flock is that lambs may not be thriving as expected. Several lambs may have a tucked-up and open fleeced appearance with a few showing faecal staining around the hindquarters due to diarrhoea (Fig. 10.3). Lambs eventually lose their appetite and become weak and unthrifty. As the disease progresses, some lambs show profuse watery diarrhoea, often containing streaks of blood. If left untreated, these animals may continue to scour and eventually die of dehydration.

On post mortem examination, the caecum is usually inflamed, empty and contracted, with a hyperaemic, oedematous and

Table 10.1 Species of coccidia recognized in sheep in the UK.

Species	Site of infection
Eimeria ovinoidalis	Ileum and caecum, colon
E. crandallis	Ileum and caecum, colon
E. bakuensis	Small intestine
E. ahsata	Small intestine
E. faurei	Small and large intestine
E. granulosa	Unknown
E. intricata	Small intestine
E. marsica	Unknown
E. pallida	Unknown
E. parva	Small intestine
E. weybridgensis	Small intestine

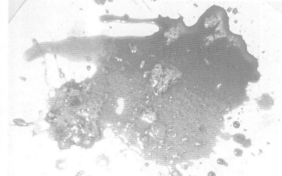

Fig. 10.3 (Left) Clinical signs of acute coccidiosis include faecal staining of the hindquarters, depression, inappetence and weight loss. (Right) Profuse blood-stained watery diarrhoea typical of acute coccidiosis.

thickened wall. In some cases the mucosa may be haemorrhagic. The ileum and colon may also be affected. Other lesions are more specific but are not usually associated with clinical signs.

With *E. ovinoidalis*, giant first-generation meronts are formed in the mucosa of the small intestine about 10 days after infection. These can cause three kinds of host reaction: leucocyte (neutrophils and eosinophils) and macrophage infiltration, crypt hyperplasia and epithelial loss. Much of the damage to the caecum is associated with the gamonts because they are the most numerous. They cause loss of surface and crypt epi-

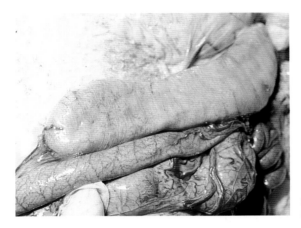

Fig. 10.4 Caecum of a lamb that died of coccidiosis.

thelium, resulting in the caecal mucosa becoming denuded (Fig. 10.4).

Eimeria bakuensis produces well-circumscribed patches of gamonts and oocysts, 1–2 mm diameter. In some circumstances, possibly owing to immunosuppression, the multiplication in these oocyst patches becomes continuous, resulting in enormous concentrations of parasites within intestinal polyps (Fig. 10.5).

Eimeria crandallis may cause wholesale destruction of infected cells by the immune system, resulting in widespread denudation. When sudden epithelial loss does occur, exudate, elec-

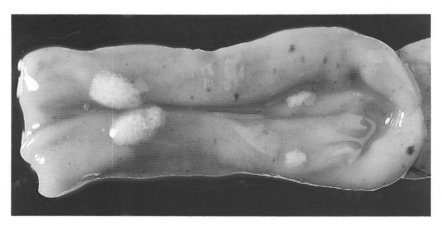

Fig. 10.5 Intestine coccidial (*Eimeria bakuensis*) polyps in the small intestine.

trolytes and proteins pour into the lumen from the denuded areas. Coagulation of these proteins sometimes forms pipe-like casts which are passed in the faeces.

EPIDEMIOLOGY

In Britain, coccidiosis most commonly affects unweaned lambs aged 4–8 weeks, either indoors or on heavily stocked pasture in cold wet weather. It is most frequently seen in the spring on lowland farms where stocking rates can be as high as 20 ewes per hectare. Stress undoubtedly plays a part in the susceptibility of lambs to coccidiosis. The disease is much less common in single lambs than in twins and triplets, which suggests that nutrition is also an important factor. On hill and upland farms, coccidiosis is much less frequently encountered, in part due to the propensity of indigenous breeds of sheep to produce single lambs, coupled with the much lower stocking rates.

The ewe is often considered to be the source of infection for lambs and initially this may be true. If infected soon after birth, a lamb's first encounter with coccidia usually causes no disease. However, the organisms can still establish themselves and multiply enormously. Thus, if a lamb picks up a few thousand oocysts during its first week of life it is likely to release several thousand million into the environment 2–3 weeks later, i.e. at about 3 weeks old. When these billions of oocysts have sporulated they then pose a tremendous challenge to lambs born later, at precisely the time of their maximum susceptibility.

DIAGNOSIS

Diagnosis should be based on history, clinical and pathological signs, as well as faecal oocyst counts and species of coccidia present. Coccidiosis should be suspected when there is severe diarrhoea (with or without blood) present in young animals, the predominant species in faecal samples is *E. ovinoidalis* and, on pathology, there is diffuse inflammation and thickening of the caecum (perhaps also of the ileum and colon, with masses of gamonts and oocysts in scrapings). It is important to note that oocyst counts alone are not a reliable indicator because not only can healthy animals pass very large numbers of oocysts, but animals can die of coccidiosis before any oocysts are shed. Furthermore, oocyst output may be transient, so an animal that is dying of coccidiosis may show very few.

DIFFERENTIAL DIAGNOSES

Diarrhoea can occur in any age of lambs up to adulthood. In newborn lambs, it is usually associated with a range of bacterial and viral pathogens, as well as cryptosporidia. There are a number of other conditions in older lambs which can be confused with coccidiosis (Table 10.2). At pasture, helminth infections, particularly *Nematodirus* species, may produce symptoms very similar to coccidiosis; concurrent infections can be particularly damaging. Lambs turned out onto recently fertilized pasture, particularly during dry weather, may develop symptoms of diarrhoea with occasional deaths. Mineral and vitamin deficiencies or poor nutrition can predispose lambs to coccidiosis, but may themselves be a cause of weight loss or poor performance.

Housed lambs on high protein creep feeds and/or milk replacements may scour and lose their bloom. Salmonellosis may also produce symptoms of diarrhoea, weight loss and sudden death. As with lambs infected with coccidia, those that do recover can take weeks to return to normal or may remain stunted permanently.

Table 10.2 Differential diagnosis of coccidiosis.

Lambs at pasture	Lambs indoors
Helminthiasis (nematodiriasis)	Salmonellosis
Poisoning (fertilizers)	High protein diet
Deficiency diseases	

PREVENTION

Lambs particularly at risk from coccidiosis are those kept indoors on damp bedding and those on contaminated heavily stocked pasture, especially during cold wet weather. The incidence of disease can be reduced through avoidance of overcrowding and stress, and attention to hygiene. Raising of food and water troughs, for example, can help avoid contamination by reducing the levels of infection. Young animals should be kept off heavily contaminated pastures when they are most susceptible. Good feeding of dams prior to parturition and creep feeding of their progeny will also help to boost resistance to coccidiosis.

Only decoquinate (Deccox; Rhône Mérieux) is licensed in the UK for the prevention of coccidiosis in lambs. Monensin has been used, although it has never been licensed as an anticoccidial for ruminants.

MANAGEMENT OF DISEASE OUTBREAKS

Outbreaks of clinical coccidiosis can appear suddenly and may prove troublesome to resolve as they often occur on heavily stocked farms, particularly where good husbandry and management are lacking. If deaths are occurring, early confirmation of the diagnosis is vital and should be based on the history, post mortem examination and examination of smears. Lambs should be medicated and moved to uncontaminated pasture as soon as possible.

Only a few licensed products are available for the treatment of coccidiosis; these can be administered either by injection or as in-feed medication. Of the sulphonamides, only sulphadimidine and sulphamethoxypyridazine are licensed for the treatment of coccidiosis in sheep; individual products can be given by subcutaneous, intramuscular or intravenous injection and sulphadimidine is also available as an oral solution. Decoquinate may alternatively be used as a creep feed additive for the treatment of clinical disease.

Normally, all lambs in a flock should be treated as even those showing no symptoms are likely to be infected. Severely affec-

M.A. Taylor

Table 10.3 Drugs available for the treatment and prevention of coccidiosis.

Chemical	Trade name(s)	Use	Dose
Decoquinate	Deccox (Rhône Mérieux)	Treatment and prevention	1 mg/kg for 28 days
Sulphadimidine	Intradine (Norbrook) Sulfoxine 333 (Vetoquinol) Vesadin (Rhône Mérieux)	Treatment	100 mg/kg then 50– 100 mg/kg daily
Sulphamethoxy-pyridazine	Midicel (Pharmacia & Upjohn) Sulphapyrine LA (Vetoquinol)	Treatment	4–4.4 ml/50 kg

ted lambs that are diarrhoeic and dehydrated may additionally require fluid therapy using either oral rehydration solutions, such as Lectade (Pfizer) or Ionalyte (Intervet), or parenteral solutions, such as compound sodium lactate (Aqupharm No. 11; Animalcare). Concurrent helminth infections, particularly with *N. battus*, may require appropriate anthelmintic treatment.

Where nonspecific symptoms of weight loss or ill-thrift are present, it is important to investigate all potential causes and seek laboratory confirmation. If coccidiosis is considered significant, much can be done through advice on management and instigation of the preventive measures outlined earlier. Batch rearing of lambs in groups of similar ages helps to limit the build up and spread of oocysts to younger crops of lambs, and allows targeting of treatment of susceptible age groups through creep feeding of lambs over the danger periods.

CHAPTER 11

Maedi-visna Virus Infection in Practice

NEIL WATT, PHIL SCOTT AND DAVID COLLIE

INTRODUCTION

Maedi-visna virus (MVV) is the prototype virus of the retroviral subfamily Lentivirinae, the members of which infect species as diverse as man (human immunodeficiency virus), primates, goats, cattle, horses and cats. In sheep, MVV causes respiratory disease (maedi), nervous disease and wasting (visna) (Fig. 11.1), mastitis and arthritis. The lesions caused by the virus comprise chronic active inflammatory changes with lymphoid infiltration and proliferation in the affected organ system. In addition to these general changes there are organ-specific changes such as smooth muscle hyperplasia in the lung, demyelination in the central nervous system, fibrosis in the mammary gland and, in the joints, proliferation of the synovial membrane and degenerative changes of the articular cartilage. The lesions can be reproduced experimentally using purified viral inocula; however, the precise mechanisms by which the virus causes damage remain unclear and are currently a major focus of research into this disease.

Fig. 11.1 A sheep with visna displaying characteristic knuckling of the fetlock joint.

THE SITUATION IN THE UK

Infection with MVV was first reported in the UK in 1979, when seropositive animals were detected in a flock in which rams were intended for export. On further investigation of the flock, approximately 60% of 185 sheep tested were seropositive for the virus. Despite this level of infection there had been no evidence of clinical disease. Subsequent studies revealed classical pathological lesions of maedi, without clinical respiratory signs, in one thin animal. In Scotland in 1981, pathological lesions characteristic of maedi were confirmed in a Texel ewe and an "in-contact" Bluefaced Leicester, but again with no clinical disease.

The first clinically patent, pathologically confirmed case of maedi was described in England in 1982. Investigation of infected flocks in East Anglia revealed histological evidence of maedi but this was often coincident with other lesions caused by pasteurellosis and sheep pulmonary adenomatosis. Prior to 1984, infection had only been reported in imported sheep and their progeny or indigenous breeds that had been in contact with imported sheep. However, during and subsequent to 1984, serological, pathological and suggestive clinical evidence of MVV infection was reported in indigenous breeds which apparently had had no contact with imported breeds. Once again, however, interpretation of the pathological findings was made difficult by the presence of other pathologies caused by *Pasteur-ella* species and *Actinomyces pyogenes*. In a follow-up to these studies, maedi coexistent with sheep pulmonary adenomatosis was reported in one of these flocks and nervous signs suggestive of visna were seen. In none of the series of investigations between 1979 and 1990 was MVV-induced mastitis or arthritis recognized in either the individuals or the flocks examined.

The degree of penetration of MVV into the national flock is not known. In one study of commercial flocks in part of East Anglia in 1987, approximately one-quarter of those tested contained infected animals. Recent evidence from a serological survey by MAFF suggested that there may be 70,000 infected sheep in the UK and that up to 1.5% of flocks may harbour infected animals. However, the survey was not designed specifically to look for MVV infection and, as only 10 animals were sampled per flock, had low sensitivity.

In 1990, a case of visna was reported in a Scottish sheep flock (Watt *et al.*, 1990) and, later, further studies were published on the clinical and pathological features of primary, uncomplicated MVV infection in this flock (Watt *et al.*, 1992). The authors' experience with this flock illustrates the spectrum of clinical problems which can occur with the infection, highlights the difficulties of diagnosis on clinical and pathological criteria alone, and indicates the possible future consequences of MVV infection in sheep flocks in the UK.

CLINICAL AND PATHOLOGICAL FEATURES

RESPIRATORY DISEASE (MAEDI)

Clinical presentation

The earliest clinical sign of maedi is exercise intolerance, manifested as increased respiratory rate and depth, abdominal expiratory effort, neck extension and open mouth breathing (Fig. 11.2). The variable response of sheep to exercise makes interpretation of early clinical signs difficult and, as a result, clinically affected animals can easily be missed on cursory examination. Thereafter, dyspnoea becomes apparent after only

Fig. 11.2 Maedi-visna virus-seropositive Texel ewe showing open mouth breathing, characteristic of maedi. Note that the animal is in good bodily condition.

mild exercise and eventually occurs in the unstressed resting animal. This is an afebrile disease. As it progresses, loss of body condition becomes apparent.

On auscultation, although breathing sounds are increased in volume, adventitious crackles and wheezes are inapparent. A dry cough is occasionally noted. In the authors' particular flock, secondary bacterial infections of the respiratory tract rarely complicate diagnosis. It has been the experience of others that this disease often occurs in conjunction with sheep pulmonary adenomatosis, pasteurellosis or chronic suppurative pneumonia.

Although lambs may become infected within the first few months of life, clinical signs do not develop until they reach adulthood (over 2–3 years old). Once clinical disease becomes apparent, sheep are not expected to live more than 8 months, the actual period being determined by the stresses to which they are exposed.

Differential diagnoses

Sheep pulmonary adenomatosis ("jaagsiekte")

This disease, caused by a retrovirus unrelated to MVV, is characterized by the development of pulmonary adenocarcinoma. Clinical signs occur in sheep over 2 years old (typically in 3- to 4-year olds) and are similar to maedi in that progressive dyspnoea and loss of body condition occur in otherwise bright animals.

In sheep pumonary adenomatosis, however, copious frothy respiratory secretions accumulate in the respiratory tract giving rise to adventitious crackles and wheezes on auscultation and these infectious fluids may pour from the nose if the hind-quarters are lifted ("wheelbarrow test"). Once clinically apparent, disease progression is rapid (a matter of weeks rather than months). There is currently no laboratory diagnostic test available to identify infected sheep.

Atypical pneumonia

This usually occurs in animals less than 1 year old. *Mycoplasma ovipneumoniae* and *Pasteurella haemolytica* are the most important organisms involved. The disease is not usually fatal, but can result in chronic ill thrift and variable degrees of dyspnoea, coughing and mucopurulent nasal discharge. A diagnosis is usually established at post mortem examination.

Chronic suppurative pneumonia

Sheep may be left with permanent lung and pleural pathology as a sequel to acute respiratory tract disease (such as pasteur-ellosis or inhalation pneumonia). Such pathology may give rise to variable functional impairment with a typical clinical presentation of variable pyrexia, ill thrift, dyspnoea and adventitious sounds on auscultation.

Lungworm

A patent infection with *Dictyocaulus filaria* will occasionally result in a persistent cough, mild dyspnoea and loss of body condition in young sheep during their first year's grazing, at the end of summer/start of autumn. Diagnosis is aided by faecal examination for lungworm larvae.

Pathological findings

On opening the thoracic cavity of severely affected sheep, the lungs do not collapse as fully or as quickly as normal lungs. Grossly, the lungs are pale, voluminous and heavy (up to four times normal weight) and there is an absence of pulmonary oedema, consolidation or collapse. Small, grey spots, 1–2 mm in diameter, may be seen subpleurally over affected lung tissue. Although there may be regional differences in severity, maedi lesions are diffuse and affect all lung lobes. On sectioning, the lungs have a firm, rubbery texture; the cut surface shows grey spots similar to those seen subpleurally and is otherwise homogeneous and slightly glassy in appearance. The regional lymph nodes draining the lung, the bronchial and caudal mediastinal nodes, are consistently enlarged, sometimes up to 15–20 times their normal size, with very prominent cortices (Fig. 11.3).

The virus induces a lymphocytic interstitial pneumonia which is characterized by a thickening of alveolar septae due to the infiltration of lymphocytes and monocytes, and the formation of lymphoid nodules with germinal centres, both in the alveolar parenchyma and around blood vessels and airways (Fig. 11.4). Most affected sheep show widespread hyperplasia of smooth muscle, particularly that associated with terminal bronchioles. In uncomplicated cases there is almost no infiltration of cells into the airspaces or airways, and fibrosis is rarely a feature of this disease. The alveolar epithelium may

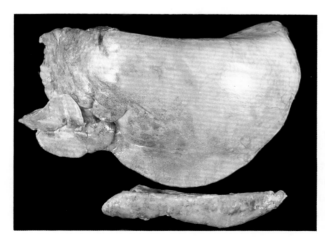

Fig. 11.3 Left lung and caudal mediastinal lymph node from a severe case of maedi. The lung is pale, voluminous and heavy (total lung weight 1.96 kg) and the lesions are uniformly distributed throughout.

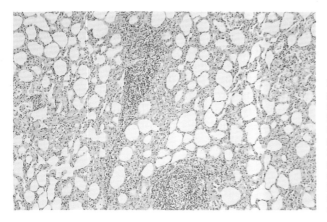

Fig. 11.4 Histological examination of the lung shows a marked interstitial reaction, with thickening of alveolar septae due to lymphocytic infiltration and hyperplasia of smooth muscle, and the formation of lymphoid follicles adjacent to blood vessels and airways.

show focal proliferation, particularly in the regions adjacent to small airways, but this is a minor lesion. Severe fibrosis, when it does occur, is usually associated with secondary bacterial infection, as are cellular infiltrates into the airspaces and airways.

Differentiating maedi-visna virus infection from sheep pulmonary adenomatosis

Sheep pulmonary adenomatosis and maedi have often been reported to occur together. The lesions of sheep pulmonary adenomatosis start as distinct nodules surrounded by normal lung tissue, but the nodules progress to involve confluent areas of the cranioventral parts of the cranial and caudal lobes, as well as mid-parts of the caudal lobes. Grossly, the lesions are firmer and more cellular than maedi lesions and often have increased amounts of clear or frothy fluid present in the airways in affected and surrounding lung. The regional lymph nodes may be enlarged up to five to 10 times their normal size, but this is not as consistent a finding as it is in maedi.

The main pathological lesion in sheep pulmonary adenomatosis infection is the proliferation of type II pneumonocytes within the alveolar epithelium, terminal bronchioles and alveolar ducts. The affected epithelium projects in papillary stalks into the airspaces which contain increased numbers of activated alveolar macrophages. Connective tissue may proliferate underlying the epithelium, in some cases producing a severe fibrosing reaction. Epithelial proliferation, marked macrophage infiltration and fibrosis are not features of uncomplicated maedi.

Secondary infection of sheep pulmonary adenomatosis lesions is common. Thus, the proliferative change in the alveolar epithelium may be complicated by an inflammatory response with large numbers of neutrophils and fibrin deposits in the airways and interlobular septae.

NERVOUS DISEASE (VISNA)

The authors have observed essentially two forms of visna: a "brain form" manifesting clinically as head tilt, circling behaviour and ataxia, and a "spinal form" which presents initially as a unilateral pelvic limb conscious proprioceptive deficit, progressing to pelvic limb paralysis. In each form of visna the neurological signs are insidious in onset, with gradual deterioration over a period of months. Therefore, it is essential to obtain a precise clinical history for each animal.

Brain form

Sheep affected by the brain form of visna have a poor bodily condition (condition score under 1.5), but remain bright and alert. The neurological signs are insidious in onset and present as a head tilt of approximately 5–10° from the vertical plane, and circling towards the affected side. These clinical signs result from lesions within the lateral ventricles. No evidence of facial nerve (VII) palsy has been observed in the visna cases in this series. In addition, some affected animals may display hypermetria and pelvic limb ataxia.

There is a slow deterioration of neurological function and affected sheep are usually destroyed on humane grounds within two months of the onset of signs, due to their poor bodily condition and hopeless prognosis.

Differential diagnoses

Vestibular lesion. A peripheral vestibular lesion causing head tilt, with or without ipsilateral facial nerve paralysis, is an important differential diagnosis of the brain form of visna. Affected animals are bright and alert with a normal appetite and there is a good response to antibiotic therapy. There may be evidence of otitis externa/media in association with the vestibular lesion, but such infections usually result from ascending infection of the eustachian tube.

Scrapie. Sheep affected with scrapie show a wide range of neurological signs, some of which – pelvic limb ataxia, hypermetria, wide-based stance with preservation of muscle strength – are observed in certain visna cases. Head tilt and circling behaviour are not observed in scrapie. In addition, scrapie-affected sheep have a vacant, detached mental state but are frequently hyperaesthetic to tactile, visual and auditory stimuli, and many exhibit a marked "nibble response" following stimulation of the skin over the sacral region.

Space-occupying brain lesions. Abscess formation or *Coenurus cerebralis* cysts, for example, generally affect one cerebral hemisphere causing contralateral blindness and proprioceptive deficits. Circling toward the affected side is commonly observed in coenuriasis.

Lumbosacral cerebrospinal fluid analysis

There is little or no change in lumbosacral cerebrospinal fluid samples collected from sheep with a peripheral vestibular lesion, scrapie or well-encapsulated brain abscesses. Cerebrospinal fluid changes in naturally occurring clinical cases of visna are mild and nonspecific, being confined to a slight elevation in protein concentration (Table 11.1). A lumbosacral cerebrospinal fluid sample will therefore be of limited use in the investigation of the brain form of visna.

Spinal cord form

The initial neurological signs of the spinal cord form of visna are hypometria and dragging of the distal limb, conscious

Table 11.1 Cerebrospinal fluid parameters (median values) in control sheep and those with either clinical maedi or visna infections.

	Maedi (*n* = 17)	Visna (*n* = 11)	Normal (*n* = 117)
Protein (g/l)	0.52 (range, 0.2–1.2)	0.76 (0.15–1.87)	0.30 (0.1–1.1)
Cells (× 10⁹/l)	0.02 (0.12–0.7)	0.012 (0.012–0.75)	0.012 (0.012–0.4)
Lymphocytes (%)	41 (12–100)	41 (13–100)	50 (0–100)
Neutrophils (%)	25 (0–70)	24 (0–71)	20 (0–84)
Histiocytes (%)	32 (0–75)	34 (0–77)	28 (0–100)

proprioceptive deficits and reduced weightbearing affecting one pelvic limb (Fig. 11.5). On cursory visual examination, these signs can easily be dismissed by a farmer or veterinarian as upper leg lameness involving either the hip or stifle joints. As the condition progresses, the dorsal surface of the hoof remains in contact with the ground when the leg is weightbearing with characteristic knuckling of the fetlock joint.

Frequently it proves difficult to assess the response of the pelvic limbs to the withdrawal and tendon jerk reflex arcs, and whether these findings indicate upper motor neuron (T2–L3) or lower motor neuron disease affecting the pelvic limbs (L3–S2).

The neurological signs gradually deteriorate over a period of 2–4 months to produce pelvic limb paresis, and affected animals frequently adopt a "dog-sitting" posture with the thoracic limbs held in extension. At this stage, sheep are destroyed on humane grounds.

Differential diagnoses

Compressive spinal cord lesions. The unilateral pelvic limb involvement in the spinal form of visna allows many spinal cord lesions, such as epidural or vertebral body compressive

Fig. 11.5 A case of visna, showing poor bodily condition, wasting of the left hindlimb and lack of proprioceptive reflexes. The animal is bright and alert and not pyrexic.

lesions which affect both pelvic limbs equally, to be excluded from the differential diagnosis list.

Joint lesions of the pelvic limb. These can be excluded by careful palpation and determination of joint excursion and crepitus. With infective joint lesions, there may be enlargement of the local drainage lymph node.

Peroneal nerve paralysis. This condition is characterized by over-extension of the hock joint and the absence of skin sensation over the craniolateral aspect of the limb distal to the stifle joint. It is uncommon in sheep.

Cerebrospinal fluid analysis

The unilateral nature of the spinal lesion is highly suggestive of visna because compressive spinal lesions almost invariably affect both pelvic limbs equally. However, it may prove difficult to assess pelvic limb reflexes, especially in large stoic sheep. Compressive inflammatory spinal cord lesions, such as epidural or vertebral body abscessation, cause protein effusion into the cerebrospinal fluid with stagnation caudal to the lesion, referred to as Froin's syndrome. With a compressive spinal lesion rostral to L5, collection of lumbosacral cerebrospinal fluid reveals a markedly raised protein concentration (commonly 2.0 to 3.0 g/l; normal range, < 0.3 g/l). The primary spinal lesion in visna is one of demyelination without cord compression. Thus, the lumbosacral cerebrospinal fluid protein concentration is within the normal range or only slightly raised. Cerebrospinal fluid analysis has proved useful in the differentiation of the spinal form of visna from other spinal cord lesions.

Pathological findings

Grossly, the brain and spinal cord are unremarkable in the majority of cases of visna. However, in animals with severe clinical signs, lesions may be visible in the wall of the lateral ventricle and/or in the dorsal columns of the spinal cord. The lesions are irregular-shaped reddish streaks (brain) or segmental lesions (cord). Consistent histological findings are a mild, nonsuppurative meningitis and focal mononuclear cell infiltration of the choroid plexus, occasionally with germinal centre formation (Fig. 11.6). In the brain and spinal cord there are multiple, focal areas of perivascular cuffing, gliosis, astrocytic hypertrophy and demyelination. These are distributed in the wall of the lateral ventricles in the dorsal and lateral columns

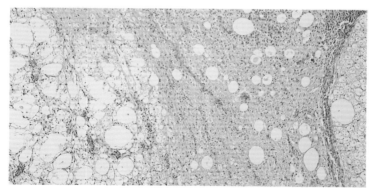

Fig. 11.6 Histological section of lumbar spinal cord from the animal in Fig. 11.5. There is demyelination and cellular infiltration in the left dorsal spinal tract, and lymphocytic infiltration of the median septum.

of the spinal cord. Gitter cells are prominent in the demyelinated areas.

MASTITIS

The clinical features of the indurative mastitis caused by MVV infection are largely unremarkable and this is rarely a major presenting sign in adults. A diffuse hardening of the mammary gland can be appreciated with experience. Discrete hardenings and variable consistency of the udder on palpation, features more indicative of chronic suppurative mastitis, are present in a proportion of MVV-infected animals but are easily distinguished from the chronic diffuse mastitis of uncomplicated MVV infection.

Grossly, the mammary glands are small and firm to cut. A diffuse lymphoid infiltration, fibrosis and the loss of acinar tissue are consistent histological changes (Fig. 11.7). In some cases, lymphoid follicles are prominent, particularly adjacent to the ductal system, and macrophages are present in the lumina of the ductal system.

In some animals, lesions of chronic suppurative mastitis are superimposed on those caused by MVV.

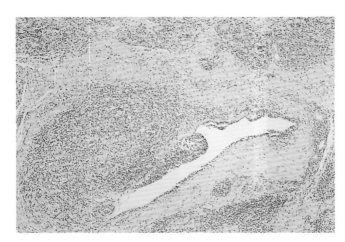

Fig. 11.7 Histology of a mastitic mammary gland shows periductal lymphoid follicle formation, lymphocytic infiltration of glandular acini and marked fibrosis.

ARTHRITIS

Clinically, arthritis is difficult to detect in most animals although some have swollen radial carpal joints (Fig. 11.8). More often the only clinical signs are a stiff, straight-legged gait, particularly noticeable after a long period of rest, and a hunch-backed stance.

In the radial carpal joints of an animal with enlarged carpal and tarsal joints the synovial membrane is red/tan-coloured and markedly thickened, and there are often erosions of the articular cartilage (Fig. 11.9). In other animals there are mild gross changes (tan-coloured synovial membrane) with no gross thickening of the membrane or joint capsule. In many cases there are no gross changes.

Pathological findings

Histologically, consistent lesions are mild to severe proliferation of the lining layer of the synovial membrane and infiltration of lymphocytes, macrophages and plasma cells around blood vessels in the synovial villi and subsynovium (Fig. 11.10).

Fig. 11.8 Arthritis principally affects the carpal joints. The animal on the left has bilaterally enlarged joints compared with the normal animal on the right.

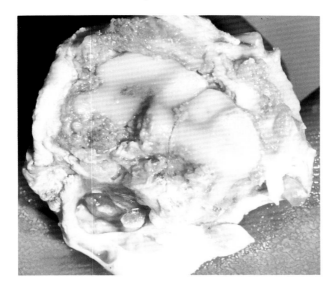

Fig. 11.9 Grossly, the carpal joints show a thickened, granular synovial membrane and erosions at the periphery of the articular cartilage.

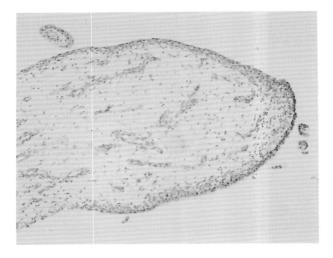

Fig. 11.10 Histologically, the synovial villi are thickened by cellular infiltration and proliferation of the synovial membrane.

Occasional additional changes include the formation of rudimentary germinal centre-like structures in the synovial membrane, fibrin deposition on synovial villi and mild neutrophil infiltration of the synovial membrane. The synovial fluid is slightly turbid in some cases.

EPIDEMIOLOGY

The transmission of MVV occurs by two main routes:

(1)   respiratory, via aerosol transmission involving respiratory secretions or, more likely, cells from the respiratory tract; and
(2)   via milk or colostrum to the offspring of infected dams.

Whether or not *in utero* transmission occurs is the subject of debate. One study failed to detect virus in 22 fetuses of infected dams, although a restricted range of tissues was examined, while another study examined fetuses from infected dams and detected virus in kidney and in lung. The authors' own studies on embryos from Texel sheep infected with a European strain of MVV failed to detect viral nucleic acids in 60 embryos from 10 infected ewes or in uterine wash fluids. Most researchers currently agree that *in utero* transmission, if it does occur, is a relatively infrequent event and is probably not epidemiologically important. The authors' findings in embryos derived from ewes infected with MVV also confirm the potential of embryo transfer techniques for the establishment of virus-free flocks and their probable safety for transmission of gene stocks between individual countries.

The time course of infection and subsequent appearance of disease in the Scottish flock studied by the authors is similar to that described during the MVV epidemic in Iceland in the 1940s and 1950s where the interval between the introduction of infection into a flock and the appearance of clinical disease was typically 8–10 years. The Scottish flock was one of the first in the UK in which infection was recognized yet clinical disease did not become apparent until 1989. The experiences with this flock indicate the danger which maedi-visna virus presents to sheep in the UK. Infection became established and widespread within the flock despite attempts, albeit not very rigorous, at control. Clinical cases of maedi, arthritis and mastitis were not recognized by the farmer; a factor in this was the insidious nature of their onset, both in individuals and at flock level, and also the difficulty of relating the variety of clinical signs to a single aetiological agent. Another factor was the perception, not unique to this farmer, that MVV does not cause clinical problems. If this pattern is followed in other flocks then we can expect the disease to become more prevalent over the next 5–10 years.

ESTABLISHING A DIAGNOSIS

The pathogenesis of MVV takes place over a protracted period and animals may be relatively old, particularly in relation to a productive lifespan, before overt clinical signs become present. This has led to the view among farmers and vets that, in commercial flocks, MVV infection may be an insignificant cause of production loss. However, experience in other countries indicates that production losses do occur, even in the absence of overt clinical disease, and may be the first presenting sign. These may be due to decreased reproductive efficiency (Dohoo *et al.*, 1987) and to lowered milk production leading to poor lamb survival and growth rate to weaning rates (van der Molen *et al.*, 1985). In the Scottish flock, lymphocytic, indurative mastitis was a marked feature of the disease and there were losses and increased culling on account of respiratory and nervous disease. The effects of these losses on the economic performance of the flock could not be assessed. However, it is likely that they were significant.

In the absence of serological investigations the disease can present problems diagnostically. The range of clinical presentations means that all clinical parameters have to be assessed, particularly those such as mastitis (which may not be immediately obvious), insidious onset nonacute lameness (which may easily be missed or masked by other lamenesses), respiratory signs (which may only be shown on exercise) and other nonspecific signs such as poor condition which may be attributed to other causes. In the authors' experience few animals die of primary MVV-induced pathology; the ones that do are usually animals with severe respiratory or central nervous system signs. Under field conditions most animals with clinical signs would probably be culled without further investigation.

If a disease problem was suspected and the animals were submitted for post mortem examination then the key to enabling a diagnosis is representative sampling of tissues, remembering that lesions may be present in lung, mammary gland, brain, spinal cord and joints. In lung, the lesions are diffuse and both ventral and dorsal and cranial and caudal parts of the lung lobes, particularly the caudal lobes, should be sampled. The relative lack of a regional distribution of lesions is a characteristic feature of the disease and a useful distinguishing feature from other ovine pneumonias. Other bacterial/viral/mycoplasmal lesions entering the lung via the airways tend to be anteroventrally distributed, blood-borne infections (e.g. bacterial emboli, parasites) tend to distribute caudodorsally or throughout the lung as discrete focal lesions, while sheep pulmonary adenomatosis tends to occur as a discrete lesion (or lesions) in the anteroventral parts of the lung or in the mid to dorsal parts of the caudal lobe.

Lung weight can be a good indicator of MVV infection, provided there is no pulmonary oedema or consolidation. In severe cases, the lung may be four times its normal weight of 0.5–0.7 kg. However, in many cases, even some with obvious clinical signs on exercise, the lung weight may only be 0.1 to 0.3 kg above normal and this can be difficult to detect, particularly if the carcase has not been bled out properly.

Histological diagnosis also presents problems as the individual lesions are not pathognomonic for the disease. Smooth muscle hypertrophy may occur as a localized lesion associated with lungworm infestation or in chronic pneumonia. Lymphoid hyperplasia occurs in mycoplasma infections and interstitial pneumonitis is associated with certain viral and chlamydial infections. However, the characteristic and distinguishing features of maedi-visna virus disease are the combination of all these individual lesions together in the lung and their distribution throughout all parts of all lobes in the lung. Their distribution is particularly important because in other sheep diseases lesions are localized either in the anteroventral parts of the lung (mycoplasmal, bacterial and viral infections) or caudodorsal parts of the lung (parasitic infestations). This emphasizes the importance of proper sampling of the lung for histopathological examination.

In the central nervous system the lesions show some similarity to *Listeria* lesions and to protozoal encephalomyelitis but can be distinguished from the former by their distribution (visna lesions are not restricted to the brain stem) and degree of demyelination, and from the latter by the lack of schizonts.

In mammary tissue not all animals show "classic" MVV-induced lesions with the formation of organized lymphoid tissue around the ductal system. In most animals there is a diffuse lymphocytic infiltrate in acinar areas and a marked degree of fibrosis. However, these lesions are not *per se* pathognomonic for MVV infection and therefore their interpretation requires some care. Again, bacteriological and mycoplasma examination has revealed no significant pathogens in these cases.

TOWARDS AN IMPROVED DIAGNOSTIC TEST

The current diagnostic test for MVV relies on detecting antibody to the virus. It involves reacting serum with a concentrated preparation of the virus in an agar gel diffusion test. The test has successfully been used to establish an MVV Accreditation Scheme which is now managed by the Scottish Agricultural College (SAC). However, it has some problems, as do all diagnostic tests. First, some animals, although infected with the virus, never produce antibody and therefore cannot be detected by this test. Secondly, the test is relatively insensitive and a high level of antibody needs to be present before it is detectable. This makes it difficult to detect sheep which have been infected recently and which have not had time to develop a high level of antibody. For these reasons, the test has to be used several times in a flock before it can be reasonably certain that all infected animals have been detected. Some of the newer technologies are now being used to develop more sensitive tests for the virus. ELISA techniques incorporating recombinant viral proteins offer possibilities for improved diagnostic tests, although their use still needs to be validated in naturally infected sheep.

QUESTIONS REMAINING

Major questions remain to be answered about maedi-visna virus. In particular we need to know:

(1) The prevalence of infection in UK pedigree and commercial flocks (SAC will soon introduce a modified scheme to assess the likelihood of there being clinically significant infection in commercial flocks)

(2)   How best to detect infected sheep, irrespective of the stage and type of infection

(3)   What effect the virus has on the economic performance of sheep, particularly their reproductive performance and ability to raise lambs, the live weight gain of fattening lambs, culling percentages, and susceptibility to other viral and bacterial infections

(4)   How the virus spreads between individuals, in particular whether there are any other routes of transmission – in utero, via bedding, faeces, people, rams, implements, etc. – which have not yet been recognized

(5)   How the virus can be eradicated from infected flocks without causing excessive financial loss.

Farmers and vets, alike, need to be alerted to the potential consequences of MVV and convinced that it is no longer an "exotic" disease but one which needs to be taken seriously in flock health schemes in this country.

REFERENCES

Dohoo, I. R., Heaney, D. P., Stevenson, R. G., Samagh, B. S. & Rhodes, C. S. (1987) The effects of maedi-visna virus infection on productivity in ewes. *Preventative Veterinary Medicine* **4**, 471–484.

van der Molen, E. J., Vecht, U. & Houwers, D. J. (1985) A chronic indurative mastitis in sheep, associated with maedi-visna virus infection. *Veterinary Quarterly* **7**, 112.

Watt, N. J., Roy, D. J., McConnell, I. & King, T. J. (1990) A case of visna in the United Kingdom. *Veterinary Record* **126**, 600–601.

Watt, N. J., King, T. J., Collie, D., MacIntyre, N., Sargan, D. & McConnell, I. (1992) Clinicopathological investigation of primary uncomplicated maedi-visna virus infection. *Veterinary Record* **131**, 455–461.

CHAPTER 12

Differential Diagnosis of Weight Loss in the Ewe

JIM HINDSON

INTRODUCTION

Weight loss in the ewe ("thin ewe syndrome") presents the practitioner with several problems in addition to that of primary diagnosis, and begs the following questions:

(1) Is the weight loss significant, i.e. does it matter?
(2) Is it true weight loss or the result of an inadequate growth rate? Does the client know which and how has he or she identified the problem?
(3) Perhaps most difficult of all, how does the veterinary surgeon inform the client that the cause may well be a function of management, in such a way as both to retain the client and provide a constructive solution?

The answer to the first question, does weight loss matter?, must certainly be an emphatic yes, in terms of both welfare and production. Even if the client is not responsive to pressure on welfare grounds, he will soon be concerned if the loss in output is quantified. An effective way of doing so is to quote the effect of body condition on ovulation rates (See Table 12.1). In other words, it pays to care.

Table 12.1 Effect of body condition score on ovulation rates in Mule ewes.

Body condition score at mating	Lambs born per 100 ewes
2	149
3	178
4	192

The question of whether the client's complaint of weight loss is, in fact, valid, requires more than subjective evidence. Most sheep keepers will be aware of the condition and wellbeing of their stock, but only rarely will they have any evidence to support a claim that either their ewes are losing weight or that the lambs are not growing at an acceptable rate. Many farmers routinely weigh lambs for selection at sale times, but very few routinely weigh their ewes.

A suggestion that the farmer identifies an indicator group in each age group on which to perform objective measurements is a good starting point for improved management. This simply requires the use of coloured ear tags. A random 5–10% of each group can then be run through a race and weigh crate at strategic points in the production cycle and the measurements recorded as weights or as correlated body score readings. If the data are available year on year, then the assessment can become an important part of "target" management.

The veterinary surgeon will need to have an idea of the target weights for the major sheep breeds in his or her area in order to be able to advise the client what a mature ewe should weigh (Table 12.2) and that, say, 250 g is a high but achievable target for daily weight gain in the commercial breeds of lamb in the post weaning finishing period. (For a discussion of the significance of body score in terms of production losses see Hindson (1989)).

Table 12.2 Pre-tupping target weights.

Breed	Weight (kg)
Suffolk	83
Texel	79
Mule/Masham	75

The degree to which management factors are likely to be involved depends on the percentage of the flock which is affected. If the problem is limited to individuals, management factors are unlikely to be of major significance; where the whole flock is affected, they assume far more importance. How, then, does the veterinary surgeon go about informing the client that the cause is probably a function of management? The first step is to avoid statements like, "Your sheep are thin because you are starving them and by the way I shall be reporting you to the DVO!" The second step is to maintain the client's self respect; this will require that the weight loss is due to some factor that he could not have anticipated or have been expected to recognize.

PROCEEDING TO A DIFFERENTIAL DIAGNOSIS

Once it is established that the client's complaint is valid, the veterinary surgeon can then proceed to a differential diagnosis.

In simple terms, weight loss is always the result of either an imbalance between supply and demand, or overt disease. In the case of the former, this may be due to a straightforward failure to provide sufficient of the major nutrients, but it can just as easily be the result of the farmer not recognizing that the flock is in a phase of increased nutritional demand (i.e. during late pregnancy or lactation). Alternatively, an animal may be failing to take in the nutrients provided (dentition problems), absorb the feed intake (chronic gut damage), or utilize the products of absorption (liver damage), or there may be protein leakage.

DEGREE OF WEIGHT LOSS

Initially, the degree and likely duration of the weight loss must be assessed. The problem may be of greater extent than the client had thought. A heavy fleece often masks the true body condition of a ewe and, as a result, weight loss may be more severe and of longer standing than would be the case in cattle or pigs. (This is one justification for winter shearing during late pregnancy.)

In human medicine, a 5% weight loss is accepted as being within the normal range of variation. In deciding what degree of weight loss is acceptable among ewes, veterinary surgeons must work from a knowledge of weights and body scores for the particular breed and the stage of the production cycle. A detailed history of any management changes must be recorded in order to correlate bodyweight loss with any such changes.

DISTRIBUTION WITHIN THE FLOCK

It is vital to establish the extent of the decline within the flock. Is the whole flock affected, only specific groups, a collection of individual cases, or an individual? If specific groups, what have they in common – age, breed or social grouping?

EVIDENCE OF CONCURRENT DISEASE

Are any of the flock:

(1) Scouring?
(2) Lame?
(3) Showing respiratory abnormalities?
(4) Anaemic?
(5) Jaundiced?
(6) Showing signs of broken fleece or cutaneous hyperaes-thesia?

If there are signs of any of the above, a detailed clinical examination of a typical case must be the first priority.

IF THE WHOLE FLOCK IS AFFECTED

DURING NORMAL/MAINTENANCE PHASE

Assuming that a problem of weight loss has been established in the whole flock, under what appear to be adequate nutritional inputs, then the quality as opposed to the quantity of the feed should be assessed (i.e. dry matter and digestibility

(D value)). A metabolic profile of a random sample of the ewe stock should be performed, which should include β-hydroxybutyrate levels and standard micronutrients. Liver enzyme levels should be checked for evidence of previous post parasitic hepatic damage, and haematology and pepsinogen levels determined to eliminate post parasitic gut damage. *Haemonchus contortus* infestation is often severe in the south and west of Britain, frequently in the absence of scouring and, because concurrent anaemia produces submandibular oedema, the condition can be confused with fascioliasis. However, the two conditions will normally occur during different seasons.

If micronutrient deficiency has been eliminated, it is highly improbable that the whole flock is affected by anything other than parasitism, which will cause a generalized weight loss. Since the removal of both the notifiable status of sheep scab and compulsory dipping, a whole flock may very rapidly become infested by *Psoroptes ovis*; if there is generalized cutaneous hyperaesthesia with fleece loss, sheep scab must be suspected.

DURING INCREASED DEMAND PHASE

The ewe's energy requirements rise to at least twice maintenance levels during late pregnancy (the last 6 weeks) and three times maintenance levels during early lactation (see Table 12.3). Competition for abdominal space in late pregnancy (when over 60% of the available abdominal volume may be occupied by the reproductive product), means that nutritional inputs must have a high D value (over 65%) and a high dry matter content (at

Table 12.3 Daily energy requirements of a ewe rearing twins.

Stage	Energy requirement (MJ)
Post weaning period	10
Nutritional flushing	15
Early pregnancy	15
Mid-pregnancy	12
Late pregnancy	20
Early lactation	25–30

N.B. In winter-shorn ewes, 10% should be added to these requirements for body temperature maintenance.

least 25%). Silage should be precision chopped to avoid intake suppression. By the time of parturition, a ewe's total daily dry matter intake will have fallen to 1.7% of its bodyweight (normal intake is 2.3% bodyweight).

There will be some loss of bodyweight in late pregnancy and early lactation in all ewes carrying multiple fetuses. The problem facing the practitioner is to establish the feed quality and ensure that the weight loss is within acceptable limits. Loss of bodyweight in late pregnancy can be masked by the rapid weight gain of the reproductive product; hence the need to check body condition scores.

Even under conditions of maintenance demand and optimum feed intake, a ewe can only improve its body condition by half a body score every 3 weeks; under high demand, recovery may be impossible. Fortunately, if corrective action is taken, ewes will not enter a prolonged "thin ewe phase" akin to the "thin sow syndrome" seen in pigs. If, however, a flock is suffering from disease in the face of inadequate nutritional inputs, the results are likely to be catastrophic, underlining the importance of eliminating concurrent overt disease.

IF SEVERAL ANIMALS OR GROUPS ARE AFFECTED

SOCIAL FACTORS

If the affected animals or group have some factor in common, such as age, breed, social or feeding group, then the differential diagnosis must first consider social factors.

Under feral conditions, sheep assume either individual range grazing or fairly small social groups. Commercial conditions, however, do not allow for these patterns. Frequently a flock will have little opportunity to establish a basic "pecking order". Under housing conditions, there will not only be social stress but often insufficient trough space to accommodate episodic feeding, even though sufficient feed is supplied.

FETAL OVERLOAD

The demand made by the fetus on a ewe's energy requirements has been discussed in relation to its effects on the whole flock.

Where several animals are affected, as opposed to the whole flock, the cause may be true "fetal overload". In very prolific breeds, such as the Cambridge or Milk sheep, multiple ovulation rates can lead to the conception of "litters" (three or more fetuses) in some individuals. Since fetal overload will occur with some degree of synchrony, it may manifest as a group problem. The end result is often pregnancy toxaemia, even in ewes receiving a full concentrate intake. Under extreme conditions, fetal overload will eventually result in recumbency.

CONCURRENT DISEASE

If social factors and fetal overload can be eliminated as possible contributing factors, then evidence of concurrent disease must be sought. By far the most common causes of group weight loss are defective dentition and lameness. While lameness is self evident, defective dentition is more difficult to identify and interpret.

Dental disease

The diseases of the sheep's mouth and teeth take two primary forms:

(1) Periodontal disease, associated with incisor wear and loss (Fig. 12.1)

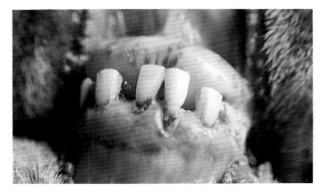

Fig. 12.1 Incisor disease.

(2)   Molar disease, leading to loss of efficiency of the cud-chewing sequence (Fig. 12.2)

The first will be quite obvious, even on the most cursory examination of the mouth. The industry associates incisor loss with loss of bodyweight and, as a result, periodontal disease is the principal reason for culling. However, records show that the correlation between body score and incisor disease is often poor and in these circumstances a veterinary surgeon may have difficulty convincing a client that factors other than periodontal disease may be involved. The correlation may, however, be

Fig. 12.2 (Top) Examination of the molar array of sheep should only be carried out with the aid of an appropriate gag. (Bottom) Post mortem examination revealing severe molar disease.

Fig. 12.3 Interdigital lesion.

greater during times of greater nutritional demand which often coincide with phases of dependence on conserved forage.

Conversely, the correlation between molar disease and body score is very high and dental surgery at the "back end" of the sheep's mouth is unrewarding. Consequently, culling should be advised where molar disease is identified.

Lameness

If dental disease has been eliminated as a differential diagnosis, and there is evidence of lameness as a flock problem, it is highly likely that pain, loss of mobility and thereby a reduction in feed

Fig. 12.4 Johne's disease, characterized clinically by chronic diarrhoea and progressive emaciation.

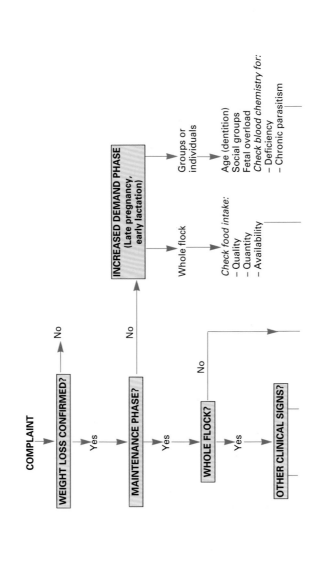

COMPLAINT

WEIGHT LOSS CONFIRMED? → No

Yes

MAINTENANCE PHASE? → No → INCREASED DEMAND PHASE (Late pregnancy, early lactation)

Yes

WHOLE FLOCK? → No

Yes

OTHER CLINICAL SIGNS?

Whole flock

Check food intake:
– Quality
– Quantity
– Availability

Groups or individuals

Age (dentition)
Social groups
Fetal overload
Check blood chemistry for:
– Deficiency
– Chronic parasitism

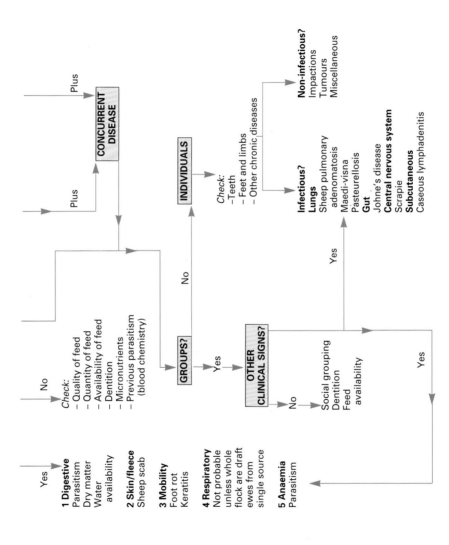

Fig. 12.5 Approach to the investigation of weight loss.

intake, are the reasons for weight loss. The degree of lameness suffered by ewes is often disproportionately greater than the severity of the lesions would suggest (see Fig. 12.3). Foot rot, of course, is not the only cause of lameness in sheep and the differential diagnoses should include other infectious causes, such as erysipelas, as well as arthritis, and spinal and joint abscessation. Impaired mobility may also be caused by visual defects which, if not identified at an early stage, can result in weight loss.

Chronic infectious disease and/or parasitism

When dental disease and impaired mobility have been eliminated, chronic infectious disease and/or group parasitism must be considered.

Often, only groups of animals will show weight loss even though the entire flock is carrying a parasite load. Consequently, the above comments relating to parasitism in the whole flock will apply here.

The chronic infectious diseases which cause weight loss include Johne's disease (Fig. 12.4), sheep pulmonary adenomatosis, maedi-visna, scrapie and chronic lung damage from previous pasteurellosis. The standard examination of lung function is not rewarding in ewes, particularly those in heavy fleece. Both the respiratory rate and body temperature fluctuate significantly with changes in the ambient temperature and humidity, and auscultation is muffled by the fleece.

IF A SINGLE ANIMAL IS AFFECTED

Where weight loss is limited to individual sheep, the clinician must resort to a detailed clinical examination (see Fig. 12.5), as the cause may be "any of those ills the sheep is heir to". It is less likely that there will be underlying faults in the management, but the cause may well be an individual case of some condition which, if not identified, will go on to affect much of the flock; maedi-visna and sheep pulmonary adenomatosis are examples of such conditions.

Molar (rather than incisor) tooth loss must be eliminated as a cause. Central nervous system involvement and cutaneous hyperaesthesia may indicate scrapie; ruminal or abomasum impaction has recently also been linked with scrapie. A ewe with chronic scouring should be examined for evidence of Johne's disease. Although polymerase chain reaction technology has greatly improved the diagnosis of this disease, the shedding of the causative agent, *Mycobacterium paratuberculosis*, is not a constant feature and, as with many sheep diseases, a post mortem examination may be the only way of obtaining a definitive diagnosis.

Diagnosis by post mortem examination is of greater significance in sheep than in other species. This is primarily the result of the "death wish" of sheep. It is also a function of the limitations of diagnosis by clinical examination and the relatively low value of individual sheep. It is a standard procedure in many parts of the world and need not be a sign of diagnostic incompetence!

ACKNOWLEDGEMENTS

The author thanks Mr T. Boundy, Dr J. Poland and Dr A. Winter for the loan of the illustrations.

FURTHER READING

Hindson, J. (1989) Examination of the sheep flock before tupping. *In Practice* **11**, 149–155.

Hindson, J. C. & Winter, A. C. (1990) *An Outline of Clinical Diagnosis in Sheep.* Blackwell Scientific Publications, Oxford.

Martin, W. B. & Aitken, I. D. (eds) (1991) *Diseases of Sheep*, 2nd edn. Blackwell Scientific Publications, Oxford.

CHAPTER 13

Caseous Lymphadenitis in Sheep and Goats

SHEELAGH LLOYD

INTRODUCTION

Caseous lymphadenitis is caused by *Corynebacterium pseudo-tuberculosis* (*C. ovis*). The encapsulated abscesses are located primarily at lymph nodes and contain material that may range from thick semi-fluid through pasty to dry and caseous or calc-ified; its colour can vary from light green through yellowish to cream. Infection is common in sheep and goats in most coun-tries of the world and prevalence rates may reach 50–60% or more. There is a high incidence of infection in some European countries, as well as in Australia and the USA from where small ruminants are imported into Britain. An infected animal does not necessarily exhibit abscesses continuously and, in 1990, caseous lymphadenitis was diagnosed in Britain in goats in con-tact with imported goats. Caseous lymphadenitis is not a noti-fiable disease and the organism is now present in several sheep flocks and goat herds in the UK. It can spread rapidly without careful diagnosis and acknowledgement of the presence of infection by herd/flock owners.

THE ORGANISM

Smears of abscess contents or culture colonies will reveal Gram-positive curved rods and cocci (Fig. 13.1). The bacteria often lie in an angular or a coryneform, palisade arrangement. Filaments of bacteria are common and many organisms are located intracellularly.

SITE OF LESIONS

The bacteria enter through cuts and abrasions in the skin and occasionally through the respiratory tract. Usually no lesion occurs at the site of entry but, after an incubation period of between a few weeks and 2–4 months, an abscess may develop in the regional draining lymph node. The entry point for bacteria can arise from several causes: wounds from shearing (compounded if the shears are contaminated with bacteria); castration, docking, etc.; and lesions from head butting and browsing, especially in goats. The abscesses become obvious in superficial lymph nodes (mandibular, parotid, retropharyngeal,

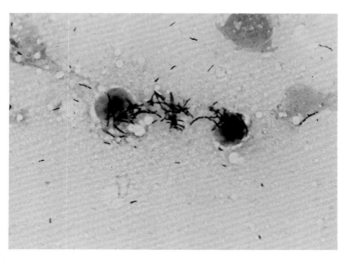

Fig. 13.1 *Corynebacterium pseudotuberculosis*; Gram-positive, mainly rods and filaments, in smear of contents from an affected goat lymph node (× 1000).

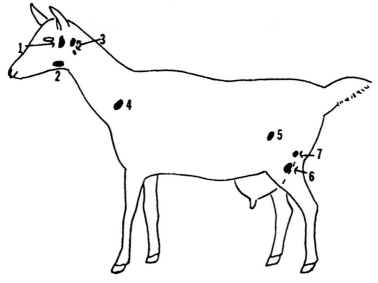

Fig. 13.2 Diagrammatic representation of lymph node sites at which caseous lymphadenitis abscesses are common and easily seen or palpated; 1, parotid; 2, mandibular; 3, retropharyngeal; 4, prescapular; 5, prefemoral; 6, mammary/superficial inguinal; 7, popliteal.

prescapular (superficial cervical), prefemoral (subiliac), mammary/superficial inguinal; Figs 13.2–13.4). An emphasis on wool production in sheep and goats' behaviour account for a greater proportion of abscesses occurring in the prescapular and prefemoral nodes and head region, respectively. Sheep only rarely have head and neck abscesses.

Fig. 13.3 Abscess in the parotid/ retropharyngeal lymph node region of a goat.

Fig. 13.4 Abscess in the prefemoral lymph node of a goat.

Visceral lesions, involving the lung parenchyma, in particular, as well as mediastinal lymph nodes and the liver, develop from the haematogenous spread of bacteria from regional lesions. Inhalation of bacteria can also be involved in the initiation of lung abscesses. Most surveys show lung and other visceral abscesses to be more common in sheep than in goats, although the age structure of population samples has differed between the two species.

At slaughter or post mortem examination, a few to 20–30 or more abscesses may be found in the lungs, with others in the liver. Abscesses are recorded in almost every organ but are conspicuous in skin, mammary glands and scrotal fascia. Mammary abscesses, because of their potential to rupture into the mammary gland, are important if milk is to be used for human consumption and could directly transfer infection to suckled young. Scrotal abscesses have a potential for, although do not usually cause, reproductive problems.

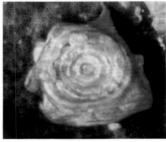

Fig. 13.5 (Top) Abscess in a sheep mediastinal lymph node becoming cream and caseous; the concentric "onion ring" configuration is developing. (Bottom) An inspissated abscess from the lung parenchyma of a sheep. The characteristic "onion ring" appearance is clearly apparent.

THE ABSCESS

Typically, an abscess may reach a diameter of 5–10 cm and is associated with a firm fibrous tissue capsule. In sheep, the contents are usually light green, becoming paler or cream to yellow as they age. Goat abscesses, more commonly although not exclusively, are cream-coloured; green is less common. The contents – at first semi-fluid yet thick (custard becoming cream cheese) – can become caseous with a firm or friable, dry texture. These contents often shell easily out of the wall. In goats, the contents often remain uniformly pasty rather than becoming dry. In sheep, the abscess, as it inspissates, develops a concentric lamellated appearance (the "onion ring" appearance; Fig. 13.5), with fibrous bands separating rings of caseous or eventually calcified material; this is unusual in goats. Sheep abscesses are also more likely to calcify than those in goats.

EPIDEMIOLOGY

An abscess contains millions of organisms which contaminate the environment on rupture. Bacteria can be found on soil, feed, feeders, fences, shears, etc. They can immediately contaminate skin cuts, for example, if shears are polluted. Dipping is a risk factor as bacteria will survive for 24 hours in the dip; this can be particularly important if dipping occurs within a week or two of shearing. Holding sheep together after shearing is another risk factor with transmission apparently from discharging lung abscesses to shearing wounds important. Bacteria potentially can survive many months in damp, shady areas in the environment.

The course of disease in a flock/herd is described in Fig. 13.6. Once the infection is endemic, abscesses will appear regularly within the flock. A wave of abscesses will occur a month or two after shearing and there are likely to be greater numbers of infected animals during and after housing. Individual animals will repeatedly develop abscesses. The prevalence of infection usually increases with age, up to about 4 years, although

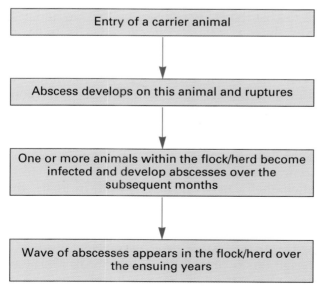

Fig. 13.6 Course of disease in a flock/herd.

immunity does gradually develop. The frequency of visceral involvement also increases with age.

DIAGNOSIS

A diagnosis of caseous lymphadenitis is dependent on there being:

(1)   A palpable abscess at a superficial or popliteal lymph node or, less commonly, on the body. The contents are cold, semi-liquid to doughy or solid. Occasionally, the abscess may be on the point of rupture;
(2)   More than one abscess, possibly a wave of them, in the herd and periodically more than one abscess on an animal. Scars or nodules from healed abscesses may be palpable at lymph nodes; and
(3)   A history of recurrent abscesses on individuals in the herd, showing increasing prevalence with age.

"Thin ewe" or "fading goat" syndromes with poor reproductive performance may be seen, while pneumonia or mastitis occur rarely.

Serological testing has been developed in the Netherlands and elsewhere. Serodiagnosis has been particularly useful in certifying flocks as caseous lymphadenitis-free (see later), from which animals can then be purchased.

LABORATORY CONFIRMATION

Confirmation of the clinical diagnosis can be obtained by culture of abscess contents.

COLLECTION OF SAMPLE

A large, 14 gauge needle aspirate of abscess contents should be collected. The thick, caseous contents can be difficult to collect, in which case repeated injection and withdrawal of a sterile bleb of air from the syringe may prove helpful. Surgical collection may be considered.

SUBMISSION OF SAMPLE

It is important to submit a brief history with the sample and to request specifically for *C. pseudotuberculosis* examination; a request which simply states abscess contents is likely to result in incomplete identification. Although *C. pseudotuberculosis* is slightly larger in Gram's stained smears (Fig. 13.1) it is morphologically similar to, and therefore difficult to differentiate from, the other corynebacteria and *Actinomyces* (*Corynebacterium*) *pyogenes*. Furthermore, a 24-h culture, as is commonly performed, often reveals little growth; what growth there is is generally very similar to the frequently encountered *A. pyogenes*. Consequently, where caseous lymphadenitis is uncommon, and even where it is common, it is often recorded as another infection, such as an unspecified coryneform or *A. pyogenes*.

FINAL CONFIRMATION

A final confirmation of identity is based on the biochemical reactions of the bacteria. A catalase test is routinely used as *C. pseudotuberculosis* is positive for catalase; it is also urease- and phospholipase-D-positive and pyrazinamidase-negative.

TREATMENT

Affected animals are not usually given systemic treatment. Although the bacteria are sensitive to many antibiotics, adequate penetration of drugs through the capsule is not practicable. An affected animal may have the abscess drained with, perhaps, local antibiotic lavage.

CONTROL

In countries where the infection is endemic, attempts may be made to control or reduce the levels of infection on a flock/herd basis. Individuals showing abscesses are isolated. Two flocks, infected and uninfected, might then be established and animals purchased only from a certified infection-free flock.

Vaccination of sheep is carried out widely in Australia. Vaccines based on partially or fully purified phospholipase-D toxoid, a virulence factor for *C. pseudotuberculosis*, do not have absolute efficacy. However, vaccination has been shown to produce a marked reduction in both the proportion of affected sheep and in the number of abscesses on individual sheep. Alternative methods of administration to produce a single-dose vaccine are under examination.

An alternative to control by vaccination is to cull infected animals. This has been developed successfully in the Netherlands. Serum is tested against concentrated exoantigen in ELISA and doubtful reactors are re-examined by immunoblot. In goats, herd specificity and sensitivity are very good but the test requires further development for sheep. Data on rate of conversion to seropositivity is needed before the test can confidently be used for identification of noninfected animals for purchase. Other tests, for example based on a recombinant phospholipase-D antigen, are under development in various countries.

With repeated testing and culling of infected individuals as they become positive (as is performed for caprine arthritis–enceaphalitis, maedi-visna, etc.), a test such as this is very useful in the control of infection within a flock. A test and cull programme, combined with hygiene, rearing kids separately without colostrum, and registration of animals, eradicated infection from > 10,000 goats on an approximately 50 farm cooperative in the Netherlands. Excellent farmer awareness and cooperation were considered essential. New goats should be purchased only from certified caseous lymphadenitis-free herds. However, a breakdown in one herd was traced to hay purchased from a farm where goats with caseous lymphadenitis were housed in the hay shed.

IMPORTANCE

SLAUGHTER LOSSES

Condemnation and downgrading of trimmed carcases result in losses and there are associated costs of inspection and removal of lesions. Hides need to be trimmed and downgraded.

MARKET RESTRICTIONS

Markets, both within countries and internationally, may be compromised. Import/export restrictions can prevent sales and/or require inspections and documentation even where the infection may already be present in the importing country/herd.

SHOW ANIMALS

Abscesses are unsightly and will restrict exhibition of valuable show animals.

PRODUCTION LOSSES

The level to which caseous lymphadenitis affects health is in dispute. Some authors consider visceral lesions to be a major factor in the "thin ewe" and "fading goat" syndromes and, hence, a factor in early culling for reasons of poor condition, poor milk and/or wool production and poor reproductive performance. Natural infection in sheep reduces wool production with an estimated annual loss of Aust$17 million to the Australian wool industry. Experimental infection with large numbers of bacteria has reduced reproductive efficiency through abortions and birth of infected lambs. Undoubtedly, some animals are clinically affected and culled early because of lesions. However, a conclusive role for caseous lymphadenitis in all animals is not proven because some infected individuals, even with 30 or 40 abscesses in the lungs, seem unaffected by the lesions.

Pneumonia associated with ruptured lung abscesses has occasionally been diagnosed. Mastitis is also rare. Abscesses in the mammary lymph nodes occasionally invade mammary tissue; *C. pseudotuberculosis* has been cultured from milk, but only rarely.

WOOL/HAIR CONTAMINATION

Wool/hair can be contaminated with abscess contents at shearing, thereby reducing its value. Other losses result from time for treatment of some ruptured lesions on the animal and attempted disinfection of shears.

PUBLIC HEALTH

C. pseudotuberculosis may infect people although the public health consequences of infections in sheep and goats are not marked. Infection may occur from continued close contact with infected animals and is an occupational hazard for farmers, abattoir workers and shearers. The ingestion of raw milk has only rarely been incriminated in granulomatous lesions in man. Granulomatous lymphadenitis or abscesses occur in superficial lymph nodes and these lesions are probably more common than recorded. Only rarely are more important abscesses reported in the liver or lungs, etc.

FURTHER READING

Dercksen, D. P., ter Laak, E. A. & Schreuder, B. E. C. (1996) Eradication programme for caseous lymphadenitis in goats in the Netherlands. *Veterinary Record* **138**, 237.

Hodgson, A. L. M., Tachedijian, M., Corner, L. A. & Radford, A. J. (1994) Protection of sheep against caseous lymphadenitis by use of a single oral dose of live recombinant *Corynebacterium pseudotuberculosis*. *Infection and Immunity* **62**, 5275–5280.

Lindsay, H. J. & Lloyd, S. (1991) Diagnosis of caseous lymphadenitis in goats. *Veterinary Record* **128**, 86.

Lloyd, S., Lindsay, H. J., Slater, J. D. & Jackson, P. G. G. (1990) *Corynebacterium pseudotuberculosis* infection (caseous lymphadenitis) in goats. *Goat Veterinary Society Journal* **11**, 55–65.

Menzies, P. I., Muckle, C. A., Hwang, Y. T. & Songer, J. G. (1994) Evaluation of an enzyme-linked immunosorbent assay using an *Escherichia coli* recombinant phospholipase D antigen for the diagnosis of *Corynebacterium pseudotuberculosis* infection. *Small Ruminant Research* **13**, 193–198.

Ter Laak, E. A., Bosch, J., Bijl, B. C. & Schreuder, B. E. C. (1992) Double-antibody sandwich enzyme-linked immunosorbent assay and immunoblot analysis for control of caseous lymphadenitis in goats and sheep. *American Journal of Veterinary Research* **53**, 1125–1132.

CHAPTER 14

Amputation of the Ovine Digit

PHIL SCOTT

INTRODUCTION

The main indication for digit amputation in the sheep is septic pedal arthritis which occurs not uncommonly, affecting individual animals on intensively managed sheep units. It results in chronic severe lameness, reduced production and rapid loss of body condition. As such, it represents a serious welfare concern. Deep penetrating wounds of the sole or deep infections secondary to solar ulceration, meanwhile, are very uncommon in sheep, although infection may result from traumatic injury – particularly if an object penetrates at the coronary band and is directed distally.

CLINICAL INDICATIONS

Clinical examination of septic pedal arthritis cases reveals severe lameness; commonly the animal fails to bear weight on the affected limb even while at rest. The pelvic limb digits are twice as likely to be affected as the forelimb digits, with medial and lateral digits equally involved. Typically, the lower limb is grossly swollen and there is obvious widening of the interdigital space. Often, there is a break in the interdigital skin and an associated purulent discharge. The coronary band of the affected digit will be markedly swollen and painful, especially on the abaxial aspect (Fig. 14.1). The skin above the coronary band shows hair loss and thinning, extending proximally for between 5 and 25 mm, and discharging sinuses with tenacious yellow/green pus, may be present. The drainage lymph node, either the prescapular or the popliteal lymph node, is three to four times the normal size and easily palpable both on account of this and the limb muscle atrophy. While detailed examination of the hoof horn at the sole/axial wall margin may reveal evidence of benign footrot, a puncture wound of the sole is rarely associated with septic pedal arthritis.

Ancillary tests, such as joint fluid analysis, are not warranted as the joint contains only a small amount of viscous pus which cannot be readily aspirated. Radiography reveals extensive soft tissue swelling and widening of the distal interphalangeal joint of the affected digit.

Separation of the white line of the abaxial wall and impaction of foreign material are common occurrences in sheep, leading to localized abscess formation. If such abscesses are not promptly drained, infection may track proximally up the hoof wall. Eventually, rupture and drainage of the abscess will occur at the coronary band and the signs of severe lameness will slowly resolve. Unlike infection of the pedal joint, wall abscesses do not result in significant inflammation of the abaxial coronary

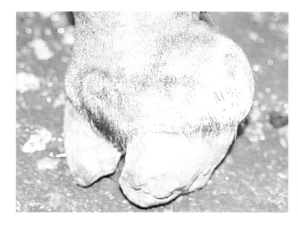

Fig. 14.1 Lateral view of the left forefoot of a Suffolk ram. Note the widening of the interdigital space and marked swelling of the coronary band with loss of hair.

band. The common association of interdigital infection with septic pedal arthritis suggests that infection gains entry to the distal interphalangeal joint axially where the joint capsule is very superficial and poorly protected. Infection tracks abaxially and proximally across the pedal joint and eventually involves the subcutaneous tissues around the coronary band of the abaxial wall.

TREATMENT OPTIONS

Prolonged antibacterial therapy is unsuccessful in treating cases of septic pedal arthritis because of the lack of effective drainage, the advanced nature of the lesion, poor antibiotic concentration achieved at the site of infection, and physical damage to joint structures. The condition is best resolved by digit amputation, although a successful method of joint lavage with hydrogen peroxide solution by means of an indwelling flushing catheter, leading to eventual arthrodesis, has been described (Corke, 1988). Successful arthrodesis of the distal interphalangeal joint has the advantage of imparting much greater foot stability but involves a more prolonged treatment course which may be limited to valuable breeding rams.

Digital amputation under intravenous regional anaesthesia (see below) provides an inexpensive method of resolving cases of septic pedal arthritis in commercial flocks. The third phalanx, navicular bone and distal one-third of the second phalanx are removed, allowing a return to full use of the limb within 4–7 days (Fig. 14.2). Although regular hoof trimming of the remaining digit is necessary, the productive lifespan following amputation remains the same as for other sheep in the flock.

SURGICAL TECHNIQUE

Intravenous regional anaesthesia is achieved by placing a tourniquet immediately above the carpal or hock joint. Between 5 and 7 ml of 2% lignocaine solution is injected through a 20 gauge 1-inch needle directed distally into a convenient superficial vein. In the pelvic limb, a suitable vein can be located

Fig. 14.2 Mule ewe 3 weeks after amputation of the medial claw of the left pelvic limb.

craniolaterally in the mid-metatarsal region. Effective analgesia is achieved within 2 min of intravenous injection and can be checked by gently pricking the skin of the coronary band of the normal digit.

The distal limb should be thoroughly cleaned, but there is no requirement for surgical preparation, as amputation will lance the subcutaneous abscesses at the coronary band causing some contamination of the wound.

A skin incision is made in the interdigital space to a depth of 1–2 mm anteriorly and going progressively deeper to 6–8 mm posteriorly (Fig. 14.3). The incision should be as close as possible to the affected digit while still removing all infected tissue.

A length of embryotomy wire is introduced through the interdigital incision and the digit removed at an angle of 15° to the horizontal plane with the abaxial side higher (Fig. 14.3). The amputation exits 8–12 mm above the coronary band which will effectively lance and drain the subcutaneous abscess(es) at this site. While a large proportion of the abscess capsule(s) will still remain, these lesions resolve once the abscess has been drained.

The second phalanx should be checked for evidence of osteomyelitis. Such infection, however, is rarely present and would be removed during amputation. Likewise, there is rarely significant infection of the deep digital flexor tendon; amputation of the digit effects good drainage of the deep digital flexor tendon sheath and no specific local treatment is required.

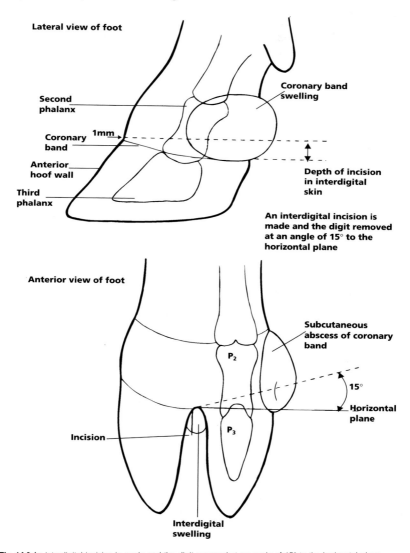

Lateral view of foot

Coronary band swelling

Second phalanx

Coronary band — 1mm

Anterior hoof wall

Third phalanx

Depth of incision in interdigital skin

An interdigital incision is made and the digit removed at an angle of 15° to the horizontal plane

Anterior view of foot

Subcutaneous abscess of coronary band

P_2

15°

Horizontal plane

Incision

P_3

Interdigital swelling

Fig. 14.3 An interdigital incision is made and the digit removed at an angle of 15° to the horizontal plane.

GENERAL ANAESTHESIA REGIMENS

In lambs of less than 30 kg, or in animals in which it is not possible to find a superficial vein for intravenous regional anaesthesia, general anaesthesia can be induced by the rapid intravenous injection of pentobarbitone sodium at a dose rate of 12 mg/kg bodyweight. This will provide surgical anaesthesia of approximately 10–15 min duration, but necessitates prior starvation for 6–12 h and close supervision until the animal is able to maintain sternal recumbency and thereby not at risk of ruminal bloat.

The recommended procedure for xylazine/ketamine anaesthesia is to administer xylazine by intramuscular injection at 0.2 mg/kg, followed by the intramuscular injection of atropine at 0.2 mg/kg. After 10–15 min, 10 mg/kg ketamine is administered by intramuscular injection inducing surgical anaesthesia within 10 min.

An alphaxalone/alphadolone combination (4 mg/kg Saffan, Mallinckrodt Veterinary) is the author's choice of injectable solution to induce general anaesthesia for digit amputation, but this product is not licensed for use in sheep in the United Kingdom. In cases where a pelvic limb digit is affected, effective caudal analgesia can be induced by lumbosacral epidural injection of lignocaine at a rate of 2 mg/kg.

POSTOPERATIVE CARE

Following digit amputation, the wound should be covered by a nonabsorbent dressing (e.g. Melolin, Smith & Nephew) and a pressure bandage with adequate cotton wool padding applied tightly. The dressing should be removed after 3–4 days by which time granulation tissue will cover much of the exposed distal portion of the second phalanx. After bandage removal, the wound should be cleaned daily with a multicleanse solution (e.g. Dermisol, Pfizer) and treated with topical oxytetracycline aerosol (see Fig. 14.2).

Postoperative analgesia is an important consideration, although currently there are no licensed analgesic preparations for use in sheep. Flunixin meglumine (2.2 mg/kg Finadyne solution, Schering-Plough) or carprofen (4 mg/kg Zenecarp injection, C-Vet) can be administered intravenously daily for 3 days, but this is only an option in hospitalized animals. Alternatively, 1 g of phenylbutazone (adult sheep) can be added to the concentrate ration daily (or drenched) for 3–6 consecutive days. Phenylbutazone is an inexpensive and effective analgesic in sheep, but unlicensed for use in this species in the UK; a 28-day meat withholding period must also be observed.

Parenteral antibiotics are not normally required following digital amputation.

PREVENTATIVE STRATEGIES

The incidence of septic pedal arthritis can be reduced by correct foot care programmes and the rapid detection and attention to infections of the interdigital space. Deep interdigital infections should be treated aggressively with an appropriate antibacterial agent, such as 44,000 i.u./kg procaine penicillin injected intramuscularly twice daily for five consecutive days.

REFERENCE

Corke, M. J. (1988) Pedal arthrodesis in sheep. *Proceedings of the Sheep Veterinary Society* **13**, 102.

CHAPTER 15

Treatment of Vaginal Prolapse in Ewes

BRIAN HOSIE

INTRODUCTION

Typically, between 1 and 2% of crossbred ewes are affected by vaginal prolapse in late pregnancy, about 20% of which die. Half of the deaths are due to rupture of the vaginal wall; the other deaths are caused by exhaustion, pregnancy toxaemia, septicaemia and toxaemia.

Most shepherds are experienced in recognizing and dealing with cases of vaginal prolapse, but there is considerable variation in their success. Veterinary surgeons in practice can offer flock owners a treatment programme tailored to the individual shepherd's ability and experience while urging them to seek professional assistance with the one or two severe cases seen in flocks each year.

Performance can be assessed at the post lambing review of the flock health programme.

TREATMENT

CLINICAL ASSESSMENT

Recently prolapsed cases are most suitable for correction by the shepherd (Fig. 15.1). In these, the vaginal wall will be moist and smooth and, while it may be swollen, congested and oedematous, it will still be warm.

Cases where the prolapsed tissue is traumatized and bleeding but still moist, congested and oedematous should be referred to a veterinary surgeon. Anaesthesia and analgesia to control straining is critical in such cases, as is antibiotic treatment to control infection.

If the prolapse is dry, cold and rough, infection, thrombosis and necrosis are likely to develop and the prognosis is poor. Shepherds should always refer such cases to their veterinary surgeon. Peracute cases of vaginal rupture where the caecum, ileum and colon herniate through a tear in the vaginal wall will die from haemorrhage and shock within an hour. Shepherds should arrange for the immediate slaughter of such animals unless a knowledge of lambing dates and the presence of col-

Fig. 15.1 A typical case of vaginal prolapse which is a good candidate for treatment.

ostrum in the udder indicate that the ewe is close to term; delivery of the lambs by caesarean section may then be possible.

ANAESTHESIA/ANALGESIA

A technique of caudal epidural anaesthesia in the ewe described by Harris (1991) produces effective anaesthesia of the vulva, vagina and perineum and minimizes straining during the subsequent manipulations. Shepherds should be urged to submit cases which will benefit from the administration of analgesia to control straining to the veterinary surgeon for treatment.

RESTRAINT

The ewe must be securely restrained and its hindquarters raised to reduce pressure on the prolapse. Where sufficient labour is available, an assistant can hold the ewe on its back with its hindquarters raised over a straw bale. Alternatively, the ewe might be strapped to a gate or, preferably, secured in a specially constructed cradle.

CLEANING AND DISINFECTION

Gross debris should be gently removed before the prolapsed tissue is washed with warm, diluted disinfectant (e.g. Cetavlon or Hibitane, ICI) and given a final rinse with warm saline or clean warm water.

REPLACEMENT OF THE VAGINA

The prolapse is reduced by pushing the periphery of the prolapse gently but firmly with bent fingers or the palm of the hand. Usually the bladder requires to be emptied by gentle manipulation and pressure or using a sharp sterile hypodermic needle. Once the bladder is empty the prolapse is more easily inverted and the tissues returned to their normal anatomical position.

PREVENTION OF RECURRENCE

With some breeds many shepherds find that tying strands of
perineal wool across the vulva is quite successful at preventing
a recurrence. Large safety pins or preferably "blanket" or "kilt"
type pins have been used to control mild cases of vaginal prolapse, but are not recommended. Plastic or alloy intravaginal
devices are widely available but these often cause vaginitis and
consequent straining. Harnesses, or ewe trusses, are often
highly effective but must be checked and adjusted at least daily.

Many shepherds stitch the lips of the vulva with two or three
double stitches of nylon tape. The stitches should be made into
the perineal skin and not the mucous membrane, as the latter
may induce further straining (Fig. 15.2). Care must be taken to
set the sutures comfortably.

Although the author has no personal experience with the
technique, veterinary surgeons may consider using Buhner's
technique of vulval closure because of claims that it offers minimal tissue reaction with good retention (see Fig. 15.2). Small
incisions are made above and below the vulva and a length of

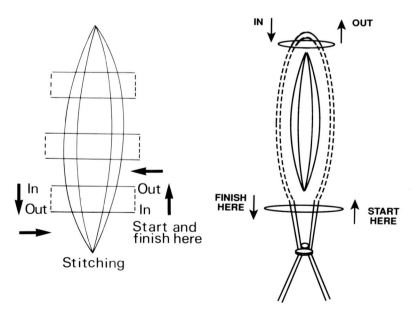

Fig. 15.2 (Left) Standard method for retaining a prolapse by stitching. (Right) Buhner's technique for retaining a prolapse.

heavy gauge nylon is laid subcutaneously, first up one side of the vulva and then down the other side, using a large, half curved, cutting suturing needle. The two ends of nylon emerging from the lower incision are tied to form a "purse string" with sufficient pressure to permit two fingers to enter the vulva. A good length of nylon must hang below the vulva to enable the shepherd to find and release the purse string at lambing.

CONTROL OF STRAINING

Nonsteroidal anti-inflammatory drugs can be used for their analgesic and antioedematous action to reduce straining.

CONTROL OF INFECTION

A parenteral long-acting antibiotic is recommended to control local and systemic infection.

IDENTIFICATION

Affected ewes must be clearly marked so that they can be assisted at lambing.

COMPLICATIONS

Continued straining, repeated prolapses, vaginitis and metritis are common consequences of poor treatment. These cases sometimes develop a secondary rectal prolapse (Fig. 15.3) but in all instances euthanasia is advisable. Exceptionally, affected ewes may be hospitalized and the pain controlled using analgesia administered by an intraepidural catheter until a caesarean section is possible.

Fig. 15.3 Rectal prolapse; the consequence of continual straining after treatment of a vaginal prolapse.

REVIEW OF RESULTS

Treatment should be reviewed if the mortality among ewes treated exceeds 10%. Because about half of ewes affected by vaginal prolapse will prolapse again in future pregnancies, affected ewes should be culled at the end of the season.

If the records show that more than 3% of the flock was affected by vaginal prolapse, attempts should be made in future years to exert greater control over nutrition of the flock in the last 8 weeks of pregnancy. A suitable control programme is described by Hosie (1989).

ACKNOWLEDGEMENTS

SAC receives financial support from the Scottish Office Agriculture and Fisheries Department.

REFERENCES

Harris, T. (1991) Caudal epidural anaesthesia in the ewe. *In Practice* **13**, 234–235.
Hosie, B. D. (1989) Vaginal prolapse and rupture in sheep. *In Practice* **11**, 215–218.

Index